Streaming, Sharing, Stealing

Streaming, Sharing, Stealing

Big Data and the Future of Entertainment

Michael D. Smith and Rahul Telang

The MIT Press
Cambridge, Massachusetts
London, England

First MIT Press paperback edition, 2017

Set in Stone Sans and Stone Serif by Toppan Best-set Premedia Limited. Printed and bound in the United States of America.

Library of Congress Cataloging-in-Publication Data

Names: Smith, Michael D., 1968- author. | Telang, Rahul, author.
Title: Streaming, sharing, stealing : big data and the future of entertainment / Michael D. Smith and Rahul Telang.
Description: Cambridge, MA : MIT Press, 2016. | Includes bibliographical references and index.
Identifiers: LCCN 2015045807 | ISBN 9780262034791 (hardcover : alk. paper) 978-0-262-53452-9 (pb.)
Subjects: LCSH: Streaming technology (Telecommunications) | Data transmission systems. | Big data. | Motion pictures.
Classification: LCC TK105.386 .S65 2016 | DDC 384.3/8—dc23 LC record available at https://lccn.loc.gov/2015045807

10 9 8 7 6 5 4

To Rhonda Smith, my best friend and the love of my life. —Michael

To my wife Ashwini and my boys Shomik and Shivum. They fill my life with so much joy. —Rahul

Contents

Acknowledgments

This book sits at the interface between two of our shared passions. First, we have a passion for great entertainment and a desire to see the motion picture, music, and publishing industries continue to be able to deliver great stories and invest in great storytellers. Second, we have a passion for using data and statistical analysis to understand how consumers and markets behave. We have many people to thank for helping us pursue both of these interests in our research and in this book.

We are indebted to our colleagues at the Heinz College at Carnegie Mellon University for allowing us to be part of a great community of scholars. In particular, we thank Dean Ramayya Krishnan for supporting our vision of a research center for entertainment analytics, and Vibhanshu Abhishek, Peter Boatwright, Brett Danaher, Pedro Ferreira, Beibei Li, and Alan Montgomery for being vital participants in that research. We also thank the many doctoral students we have had the pleasure of working with at Carnegie Mellon, including Uttara Ananthakrishnan, Daegon Cho, Samita Dhanasobhon, Anindya Ghose, Jing Gong, Anuj Kumar, Liron Sivan, and Liye Ma. Members of the staff at Carnegie Mellon have created a wonderful environment within which we can teach and do research, and we thank Andy Wasser and Sean Beggs in the Masters of Information Systems Management program, Brenda Peyser in the Masters of Public Policy and Management program, and John Tarnoff and Dan Green in the Masters of Entertainment Industry Management program for their hard work in each of these academic programs. Mary Beth Shaw in Carnegie Mellon's General Counsel Office has been an outstanding advocate for our research, and we are grateful for her help. Our colleague Millie Myers provided us with outstanding insight into external communication through her media training program. We also thank the many students we have had a chance to interact with during our time at Carnegie Mellon. In particular, we thank Chris Pope, Ricardo Guizado, and Jose Eduardo Oros Chavarria for their outstanding data analytics support.

We are thankful to the many people in the entertainment industry who have shared their expertise and experiences with us. Among others, we would like to thank Cary Sherman and his team at the Recording Industry Association of America for excellent guidance and insight into the music industry, and Al Greco and his team at the Book Industry Study Group for data and expertise pertaining to the publishing industry. We would also like to thank the Motion Picture Association of America for its continued, stalwart support of our research at Carnegie Mellon University through Carnegie Mellon's Initiative for Digital Entertainment Analytics.

We are indebted to several people for support and encouragement. Andrew McAfee encouraged us to pursue our vision and introduced us to his outstanding literary agent, Rafe Sagalyn. Rafe has provided invaluable help in crafting our vision and guiding our book in the marketplace. Jane MacDonald and her team at the MIT Press have been a delight to work with, and we are very thankful to them for taking a chance on two first-time authors. We also thank Natalie Madaj and David Israelite at the National Music Publishers Association for their help in securing copyright permissions. Finally, and most importantly, we would have never finished this book without the guidance, help, patience, and good humor of our editor, Toby Lester. Toby has the uncanny skill and ability to take our random thoughts and ideas and to turn them into exactly what we wanted to say. Without him we would still be re-wording chapter 2.

Personal Acknowledgments

I would like to thank Erik Brynjolfsson for being my advisor, coach, and mentor at MIT. I couldn't have asked for a better example of what it means to be a scholar. None of this would have been possible without the love and support of my dear wife. Rhonda, thank you for believing in me and for encouraging me to try so many things I didn't think I could do. Thank you Davis, Cole, and Molly, for the joy you've brought into our lives. Mom and Dad, thank you for your patience and for giving me a love of learning. And thank you to Jesus Christ for cancelling a debt I owe with a sacrifice I can never repay. —M.D.S.

I would like to thank my mother and father, who always believed in me and let me chase my dreams. My wife Ashwini, who is a constant source of inspiration and makes me try harder. My boys Shomik and Shivum constantly encourage me without saying a word because they believe in me more than anyone else. Finally, I am thankful to all my mentors, colleagues, and students, who teach me something new every day. —R.T.

I Good Times, Bad Times

In the days of my youth I was told what it means to be a man / Now I've reached that age, I've tried to do all those things the best I can / No matter how I try I find my way into the same old jam

Led Zeppelin, "Good Times, Bad Times"

1 House of Cards

Every kitten grows up to be a cat. They seem so harmless at first—small, quiet, lapping up their saucer of milk. But once their claws get long enough, they draw blood, sometimes from the hand that feeds them.

Frank Underwood, in the Netflix original series *House of Cards*

For the creative industries—music, film, and publishing—these are the best of times and the worst of times. New technologies have provided self-published authors, independent musicians, and other previously disenfranchised creators with powerful new ways of doing their work and reaching their audiences, and have provided consumers with a wealth of new entertainment options. Together these changes have produced a new golden age of creativity. But the same technologies also have changed the competitive landscape, weakened the control that established players can exert over both content and consumers, and forced business leaders to make difficult tradeoffs between old business models and new business opportunities. In the face of these changes, many powerful firms have stumbled and lost ground in markets they used to dominate.

One of the most profound examples of this shift in market power occurred when Netflix began to offer original programming. It's a fascinating case that illustrates many of the ways in which technology is changing the entertainment marketplace.

The story begins in February of 2011, when Mordecai Wiczyk and Asif Satchu, the co-founders of Media Rights Capital (MRC), were pitching a new television series, *House of Cards*, to several major television networks. Inspired by a BBC miniseries of the same name, the proposed series—a political drama—had attracted top talent, including the acclaimed director David Fincher, the Academy Award–nominated writer Beau Willimon, and the Academy Award–winning actor Kevin Spacey. While shopping the broadcast rights to HBO, Showtime, and AMC, Wiczyk and Satchu approached Netflix about securing streaming rights to the show after it had finished its television run.[1]

In its pitches to the television networks, MRC had focused almost exclusively on the draft script for the pilot episode and on the show's overall story arc. The goal of these meetings was to secure a commitment from a network to fund a pilot episode. The challenge involved rising above the hundreds of other creators who were pitching their own ideas, competing for the small number of programming slots owned by the major networks. But that's just how the business worked—the networks called the shots. "We had a monopoly," Kevin Reilly, a former chairman of entertainment at the Fox network, has said. "If you wanted to do television, you were coming to network television first."[2]

Pilot episodes are the standard tool that television networks use to determine whether there is an audience for a show. Creating a pilot episode requires the writers to introduce and develop the show's characters, plot elements, and story arc in a 30- or 60-minute broadcast time slot. That's difficult under the best of circumstances, but it was particularly difficult in the case of *House of Cards*. "We wanted to start to tell a story that would take a long time to tell," Kevin Spacey said in 2013. "We were creating a sophisticated, multi-layered story with complex characters who would reveal themselves over time, and relationships that would take space to play out."

Even if a proposed show receives funding for a pilot episode, the funding comes with no guarantees to the show's creator—the network is still in complete control. If the network likes the pilot, it might initially order from six to twelve episodes, but that's rare. Usually the network decides to pass after seeing the pilot, and the creators have to start over.

For the networks, pilot episodes are an expensive way to gauge audience interest. Making a pilot episode for a drama series can cost between $5 million and $6 million,[3] and some in the industry estimate that $800 million is spent annually on failed pilots—that is, pilot episodes that never lead to series.[4]

Before their meeting with Netflix executives, Wiczyk and Satchu had gotten a mixed reaction from the television networks to their pitches for *House of Cards*. The networks had liked the concept and the talent attached to the project, but no network had been willing to fund a pilot episode, in part because the conventional wisdom in the industry— since no political drama had succeeded since the final episode of *The West Wing*, in 2006—was that political dramas wouldn't "sell."[5]

The reception at Netflix was different, however. Ted Sarandos, Netflix's Chief Content Officer, wasn't primarily interested in critiquing the show's story arc or invoking the conventional wisdom about the market's taste for political dramas. Instead, he came to the meeting primarily interested in data—his data—on the individual viewing habits of Netflix's 33 million subscribers. His analysis showed that a large number of subscribers were fans of movies directed by David Fincher and movies starring Spacey. The data also revealed that a large number of customers had rented DVD copies of the original BBC series. In short, the data showed Sarandos that the show would work[6] and convinced him to make an offer to license the show directly to Netflix,[7] bypassing the television broadcast window entirely.

But Netflix's innovative approach didn't stop there. Netflix didn't make the typical offer of $5 million or $6 million to produce a pilot episode that it might option into a half-season or full-season order. Instead, Netflix offered $100 million for an up-front commitment to a full two-season slate of 26 episodes. Netflix argued that it didn't have to go through the standard pilot process, because it already knew from its data that there was an audience for *House of Cards*—and that it had a way to target potential members of that audience as individuals.

Netflix's decision not to use a pilot episode to test the *House of Cards* concept garnered a skeptical response from the television industry. In

March of 2011, shortly after the *House of Cards* deal was announced, Maureen Ryan, writing for the online service AOL TV, made a list of reasons to doubt that *House of Cards* would be successful if delivered by Netflix. Her article closed with the following observation:

> The other red flags here? Netflix and MRC are going forward with this project without stopping to make a pilot first, and Fincher's never worked on a scripted drama before. We all like to make fun of TV suits, but sometimes those suits know what they're talking about. Many pilots in TV history have been tweaked quite a bit to make them better—in some cases, a lot better.[8]

The decision to bypass a pilot episode wasn't the only difference between Netflix's approach and that of the "suits." Instead of following the traditional broadcast model of releasing one episode per week to build an audience, Netflix planned to release all of season one's thirteen episodes at once. This was unheard of in the television industry. Television broadcasters are limited to a common broadcast schedule that must meet the needs of all their viewers, and a 13-hour show would crowd out all of the network's other programming for a day. Netflix had a clear advantage over the broadcasters: Its streaming platform didn't restrict viewers to watching specific episodes at specific times. Rather, they could watch episodes at their convenience, or even "binge watch" the entire season, as 670,000 people reportedly did with the second season of *House of Cards*.[9] They also didn't have to put up with the annoyance of commercial breaks, having paid, through their subscription fee, for the right to watch the show without them.[10]

In addition to opening up new opportunities and new flexibility for viewers, the "all-at-once" release strategy for *House of Cards* opened up new creative opportunities and flexibility for Beau Willimon, the show's head writer. When writing a typical weekly series, he would have to fit each week's story into precise 22- or 44-minute chunks, depending on whether the show would be broadcast in a 30-minute or a 60-minute slot. Then, within these slots, he would have to build in time at the beginning of each episode to allow viewers to catch up with plot elements that they might have missed or forgotten, time in the middle of episodes for act breaks to accommodate commercials (the main source of revenue for broadcast content), and time at the end of episodes for "mini-cliff-hangers" to build interest for the next week's episode. In an all-at once release, none of these things were necessary, so Willimon was free to focus his energies on creating what he has called "a 13-hour movie."[11]

Knowing that they had an up-front commitment to a two-season deal, instead of the typical 6- or 12-episode deal, also helped the writers by giving them more time to develop their story. "When they opened the writer's room, they knew there was going to be a 26-hour [show], and they wrote accordingly," Sarandos said in a 2013 interview with *The Hollywood Reporter*.[12] "I think we gave the writers a different creative playground to work in, and the show is better because of it."

Netflix's subscription-based business model and on-demand content provided creative freedom for the writers in other areas as well. For example, Beau Willimon's script for *House of Cards* began by having Frank Underwood, the show's lead character, strangle his neighbors' injured dog—a scene that made a number of TV veterans at Netflix uncomfortable. "Early on," Willimon observed at the 2014 Aspen Ideas Festival, "there were a few people ... who said, 'You can't kill a dog, you'll lose half your viewership in the first 30 seconds.' So I go to Fincher and I say, 'Hey, man, I'm really into this opening. I think it really works for the opening of the show. People are telling me we'll lose half of our viewers when we kill this dog. What do you think about that?' And he thinks for a second and goes, 'I don't give a shit.' And I go, 'I don't either.' And he says 'Let's do it.'"[13]

For most television shows, that sort of creative freedom would have been almost unthinkable. In the same Aspen Ideas Forum panel, the industry veteran Michael Eisner noted that if he had tried to include a similarly violent scene in an episode for broadcast television "the president [of the network] would call me, the chairman of the board would call me, I would be out in 10 minutes."

Why would this scene work for Netflix but not for broadcast television? First, Netflix wasn't pursuing an advertising-supported business model, so it didn't have to worry about offending its advertisers by including a controversial scene. Second, because Netflix provided an on-demand streaming platform with many different options, it could risk offending individual subscribers with the content in some of those options. In a broadcast world, you can deliver only one show at a time to your audience, so that show must appeal to as many viewers as possible. But a Netflix subscriber who was repulsed by Frank Underwood's actions could choose from more than 100,000 hours of other Netflix content. In fact, by observing how individual viewers responded to this scene, Netflix was able to gather important information about their preferences. As Willimon said, "if you weren't going to be able to survive this dog strangling, this probably wasn't the show for you."

Customer data, and the ability to personalize the Netflix experience for its subscribers also gave Netflix new options to promote its shows. Incumbent television networks know the general characteristics of viewers from Nielsen estimates and other surveys, but they rarely know who their viewers are as individuals; even if they do, there is no easy way for them to promote content directly to those consumers. Typically, the best they can do for a new show is promote it alongside a similar established show, in the hopes that viewers of the latter will be interested in the former. Netflix, because it knew its customers as individuals, was able to do much more with *House of Cards*. It could see what each subscriber had viewed, when, how long, and on what device, and could target individual subscribers on the basis of their actual viewing habits. Netflix even created multiple "trailers"[14] for the show. One featured Kevin Spacey (for subscribers who had liked Spacey's movies); another featured the show's female characters (for subscribers who liked movies with strong female leads); yet another focused on the cinematic nuances of the show (for subscribers who had liked Fincher's movies).[15]

While Netflix was working hard to expand the use of digital channels to distribute and promote content, the networks were trying to find ways to *limit* the use of digital channels to avoid cannibalizing viewing (and advertising revenue) on their broadcast channels. Some people at the major TV studios understandably saw new digital channels as a threat to their current revenue streams and judiciously avoided licensing content for digital delivery. It's hard to fault them for that choice—killing the golden goose is a good way to get fired in any business.

When shows *were* licensed on digital channels, they were typically delayed by 1–4 days after the television broadcast to avoid cannibalizing "live" viewership. This followed a standard practice in the creative industries: delaying the availability or degrading the quality and usability of "low-value" products (e.g., paperback books and DVD rentals) to protect revenue from "high-value" products (hardcover books, Blu-ray discs). The practice made sense—in an à la carte business model, price discrimination is the most economically efficient way for creators to sell content.

However, in order for price discrimination to work effectively, you must be able to control the availability, quality, and usability of how customers access content. In the analog era, creators had at least a fighting chance of maintaining such control. In the digital era, control is much more difficult to exert. Now, for example, instead of having to choose between watching a network's live broadcast via a "high-value" television platform or waiting 1–4 days to watch its digital version via a "low-value" platform, digital consumers have an alluring new option: a "no-value" (to the network) pirated copy that costs nothing, has no commercials, and could be watched in high definition almost immediately after the initial broadcast. In view of this allure, it isn't surprising that traffic from the popular file-sharing protocol BitTorrent accounted for 31 percent of all North American Internet traffic during peak-traffic periods in 2008.[16]

Piracy poses an even greater risk abroad, where a television show can be delayed by several months after its initial broadcast in the United States. These delays are driven by business processes that worked well in a world in which most promotional messages were local and in which international consumers had no other options to view programs. But if you live in Sweden, and your Facebook friends in the United States are talking about the new episode of *Under the Dome*, it's hard to wait two months[17] for that show to be broadcast on your local television station, particularly when you know that it's readily available on piracy networks today.

One way to compete with piracy is by making pirated content harder to find and more legally risky to consume. To do this, studios must send out thousands of notices to search engines and pirate sites asking that their content be removed from webpages and search results. This strategy can be effective, but it requires constant effort and vigilance that some have compared to a non-stop game of *Whac-a-Mole*.[18]

Netflix, however, was able to pursue a fundamentally different strategy for distributing *House of Cards*. The company's business model was based on selling access to a bundled platform of on-demand content. Large-scale bundling was impractical for most physical goods, because of the manufacturing costs required for the individual products. But digitization eliminated manufacturing costs, making large-scale bundling of motion-picture content possible—more than merely possible, in fact: economic research has shown that large-scale bundling can generate more profit for the seller than can be generated with à la carte business models.[19]

Bundling also enables sellers to focus on new ways of delivering value to consumers. Price-discrimination strategies rely on reducing the attractiveness of some products enough that they appeal only to low-value consumers—something Reed Hastings, the CEO of Netflix, has referred to as "managed dissatisfaction."[20] In place of this managed dissatisfaction, Netflix was able to focus on convenience and accessibility: subscribers in all of the 41 countries the company served in 2013 could watch *House of Cards*, or any other Netflix program, using a single easy-to-use platform on any of their enabled devices without worrying about the legal, moral, or technical risks of piracy. Netflix would even keep track of where users were in an episode so they could pick up the series at the same spot if they needed to pause watching or switch devices. By delivering more value from their service than consumers could receive from pirated content, and by charging a reasonable fee for this extra value, Netflix hoped that most customers would find their streaming channel more valuable than what they could find through piracy. And on the surface, this strategy seems to be working. In 2011, Netflix's share of peak Internet traffic exceeded BitTorrent's for the first time, with Netflix at 22.2 percent of all North American Internet traffic and BitTorrent at 21.6 percent.[21] By 2015 the gap had widened, with Netflix at 36.5 percent and BitTorrent at only 6.3 percent.[22]

In short, Netflix's platform and business model gave it several distinct advantages over incumbent studios and networks:

• a new way to green-light content (through detailed observations of audience behavior rather than expensive pilot episodes)
• a new way to distribute that content (through personalized channels rather than broadcast channels)
• a new way to promote content (through personalized promotional messages based on individual preferences)
• a new and less restrictive approach to developing content (by removing the constraints of advertising breaks and 30- or 60-minute broadcast slots)
• a new level of creative freedom for writers (from on-demand content that can meet the needs of a specific audience)
• a new way to compete with piracy (by focusing on audience convenience as opposed to control)
• a new and more economically efficient way to monetize content (through an on-demand bundled service, as opposed to à la carte sales).

Perhaps this all means that Netflix will be the "winner" in digital motion-picture delivery. But perhaps not. Netflix, after all, faces challenges from Google, Amazon, and Apple, which, by virtue of their existing businesses, have competitive advantages of their own: the ability to subsidize content to obtain data on customers, enhance customers' loyalty, or sell hardware. Netflix also faces challenges from the studios themselves, which are using platforms such as Hulu.com to vertically integrate into the digital distribution market.

We don't want to prognosticate in this book. We don't know which firms are going to come out on top in the next phase of competition in the entertainment industries. But we *do* know how technology is changing the entertainment industries. That's because for the past ten years, as faculty members at Carnegie Mellon University's Heinz College, we have led an in-depth research program to analyze the impact of technology on entertainment. We have worked with many talented people at leading motion-picture studios, music labels, and publishing houses to use data and advanced statistical analysis to understand how technology is changing specific aspects of their business. Our research with these firms has addressed every major consumption channel—legal or illegal, digital or physical—and has touched on nearly every major marketing and strategic choice facing these industries. We have learned an extraordinary amount. Our research has yielded new insights into the business and public-policy questions facing the copyright industries, unique access to industry leaders and datasets that have helped us address those questions, and an understanding of the challenges that companies in the entertainment industries face and the business strategies they can use to overcome them.

But while we were studying these specific questions, we began to ask a more general question: Is technology changing overall market power in the entertainment industries?

From a historical perspective, the answer to this question appears to be No. For 100 years, market power in the entertainment industries has remained concentrated in the hands of three to six publishing houses, music labels, and motion-picture studios. And these "majors" have been able to maintain their market power despite extensive shifts in how content is created, distributed, and consumed. In the twentieth century, low-cost paperback printing, word-processing and desktop publishing software, recording to magnetic tape (and later to videocassettes, CDs, and DVDs), radio, television, cinema multiplexes, the Walkman, cable television, and a host of other innovations were introduced. Through it all, three to six firms—often the same three to six firms—maintained control over their industries.

The key to the majors' dominance has been their ability to use economies of scale to give themselves a natural competitive advantage over smaller firms in the fight for scarce resources. Through these economies of scale, the "majors" successfully controlled access to promotion and distribution channels, managed the technical and financial resources necessary to create content, and developed business models that allowed them to determine how, when, and in what format consumers were able to access content.

Because these market characteristics persisted throughout the twentieth century, it is natural to conclude that no single change in computing or communications technologies would affect market power in the entertainment industries. But what if the entertainment industries are facing *multiple* changes? What if advances in computing and communications technologies have introduced a set of concurrent changes that together are fundamentally altering the nature of scarcity—and therefore the nature of market power and economic profit—in the entertainment industries? Consider the following changes that have been introduced by digital technologies:

• the development of digital distribution channels with nearly unlimited capacity, which shifted the entertainment industries away from a world in which content was distributed through scarce broadcast slots and scarce physical shelf-space

• the introduction of global digital piracy networks, which make it harder for content producers to generate profit by creating artificial scarcity in how, when, and in what format consumers are able to access entertainment content

• the availability of low-cost production technologies, which shifted the entertainment industries away from a world in which only a privileged few were able to access the scarce financial and technological resources necessary to create content for mass consumption—a shift that has resulted in an explosion of new content and new creative voices

• the introduction of new powerful distributors (Amazon, Apple, Netflix, YouTube) that can use their unlimited "shelf space" to

distribute this newly available content, and which are using a new set of economies of scale to achieve global dominance in markets for content distribution
• the development of advanced computing and storage facilities, which enables these powerful distributors to use their platforms to collect, store, and analyze highly detailed information about the behavior and preferences of individual customers, and to use this data to manage a newly important scarce resource: customers' attention.

Although a variety of experts have discussed various individual changes in the creative industries, no one has looked at them as a whole or used data to evaluate their combined effects rigorously. That's what we hope to do in this book. And what we think you'll see when you look at these changes as a whole, in light of the empirical evidence, is a converging set of technological and economic changes that together are altering the nature of scarcity in these markets, and therefore threatening to shift the foundations of power and profit in these important industries. That shift, in fact, has already begun.

• • •

This is an issue that affects us all. If you are a leader in the motion-picture industry, the music industry, or the publishing industry, you may wonder how these changes will affect your business, and how your company can respond. If you are a policy maker, you may wonder how these changes will affect society, and how government can ensure the continued vitality of these culturally important industries. If you are a consumer of entertainment, you may wonder how technology will change what content is produced in the market and how you will access that content. This book provides answers to all these questions. Drawing on our access to market data and our knowledge of the entertainment industries, it integrates our findings and sums up ten years of research. It analyzes how technology is changing the market for creative content, and why—right now, in fundamental ways—these changes threaten the business models that have governed the entertainment industries for 100 years. And it proposes practical ways in which major publishers, music labels, and studios can respond.

We hope you caught the end of that last sentence. Many pundits argue, sometimes with glee, that content creators and markets for entertainment are doomed because of how technology is changing the nature of scarcity in entertainment. We strongly disagree. On the basis of our research, we are optimistic about the future health of markets for creative content. Information technology makes some business models less profitable, of course; but it also makes possible new degrees of personalization, customization, variety, and convenience, and in doing so it introduces new ways to deliver value to consumers, and new ways to profit from delivering this value.

But you can't effectively pursue these new opportunities unless you understand the historical sources of market power and economic profit in the entertainment industries. In the next chapter we'll address two foundational questions: Why do markets for creative content look the way they do? What factors have allowed a small number of firms to dominate these industries?

2 Back in Time

Don't bet your future on one roll of the dice / Better remember lightning never strikes twice
Huey Lewis and the News, "Back in Time"

Not long ago, a leader in one of the entertainment industries delivered a talk to our class. He provided us with valuable perspective on the nature of his business and the challenges it faces today, but at a certain point he said something that gave us pause. We were discussing the rise of the Internet and its effects on his industry, and someone asked if the Internet might threaten the market power of the small group of "majors" that had ruled the business for decades. Our guest speaker dismissed the question. "The original players in this industry have been around for the last 100 years," he said, "and there's a reason for that." We found that remark understandable but profoundly revealing. It was understandable for the simple reason that it was true, and we had heard other executives express nearly identical sentiments about their own industries. But we found it revealing because it didn't acknowledge something else that was true: that the technological changes at work in the creative industries today are fundamentally different from those that came before them. These changes threaten the established structure in the entertainment industries, and leaders of those industries must understand these changes and engage with them if they want their businesses to continue to thrive.

Before considering how things have changed, let's explore the market realities behind the statement the industry leader made in our class.

Why is it that in the entertainment industries so much power is concentrated in the hands of so few companies? What are the economic
characteristics that allow large labels, studios, and publishers to dominate their smaller rivals? And why have these characteristics persisted
despite regular and major changes in the technologies for creating, promoting, and distributing entertainment?

For variety's sake, because we discussed the motion-picture industry
in chapter 1, we'll focus on the music industry in this chapter,[1] with the
understanding that each of the creative industries experienced a similar
evolution. Our motivation here is fundamental. To understand how
technology may disrupt the creative industries in the twenty-first
century, we need to understand how those industries evolved in the
twentieth century.

• • •

Until the late 1800s, the music industry was primarily the music-*publishing* industry. Sheet music, copyrighted and printed and distributed
in the fashion of books, was what you bought if you liked a song and
wanted to hear it at home. You got your music in a store or at a concession stand, played it on the piano in your parlor, and Presto!—you had
a home entertainment system. New York, especially the district of Man-
hattan known as Tin Pan Alley, became the hub of the sheet-music
business. By the end of the nineteenth century, thanks to the growth of
the middle class, sales of sheet music were booming. In 1892, one song
alone, Charles K. Harris' "After the Ball,"[2] sold 2 million copies. To meet
the growing demand, music publishers signed up writers who could
produce catchy, easy-to-play songs with broad popular appeal. The path
ahead seemed clear.

Change was brewing, however. Back in 1877, while tinkering with
improvements to the telegraph, the young inventor Thomas Edison
had created a device that could record, store, and play back sound.
It consisted of a mouth horn, a diaphragm, a stylus, and a cylinder
wrapped in tin foil, and its operation was simple: To record your voice,

you spoke into the horn while turning the cylinder with a hand crank. The sound of your voice made the diaphragm vibrate, which in turn caused the stylus to imprint indentations in the foil. Playback involved reversing the process: You put the stylus at the beginning of the indentations in the foil and began rotating the cylinder. The movements of the stylus along the indentations in the foil caused the diaphragm to vibrate and make sound, which the horn amplified. Faintly, and a little eerily, your voice would re-emerge from the horn, as though overheard through a wall. Edison patented the idea immediately, using the name "phonograph." But as is so often the case with new technologies, he didn't fully recognize its potential. The quality of his recordings was poor, and they had to be produced individually, and thus his phonograph was little more than a novelty item. Within a year, Edison had moved on to a different novelty item: the electric light.

But others continued to play with the idea. In 1885, a patent was issued for a competing device called the "graphophone," which used wax cylinders rather than tin foil to make recordings. This drew Edison back into the game, and in 1888 he devised what he called an "improved phonograph," which also used wax cylinders. Not long afterward, a wealthy businessman bought the rights to both devices and formed the North American Phonograph Company. His business plan was to sell the devices as dictation machines for use in offices. That plan failed, and soon the North American Phonograph Company faced bankruptcy. Smelling opportunity, Edison bought back the rights to the phonograph, and eventually a way was found to make money from such machines by installing them in coin-operated "jukeboxes" for use in amusement parlors.

In 1889, both Edison and a new jukebox producer called the Columbia Phonograph Company began to sell music recorded on wax cylinders, and the music-*recording* industry was born. But change was brewing again. That same year, a different kind of recording device, known as the "gramophone," appeared on the market. Invented by Emile Berliner and patented in 1887, it recorded sound by means of a vibrating stylus, just as the phonograph and the graphophone did.

Instead of cylinders, however, it used flat, easily copied discs, or "records." Berliner produced his first records in 1889, for a toy store. In the mid 1890s it began to offer gramophones and records to the general public, in direct competition with the phonographs and cylinders produced by Edison and Columbia. Because records could be mass produced and stored more easily than cylinders, they had distinct advantages in that competition, and it soon became clear that they would become the industry standard. A high-stakes legal battle ensued, Columbia arguing that Berliner had infringed on its patents with his gramophone. In 1901 a judge ruled that both companies would be allowed to make records, a ruling that was deemed a victory for Berliner. To commemorate that victory, Berliner and others formed the Victor Talking Machine Company.

Victor and Columbia soon came to dominate the industry. Edison misguidedly stuck with cylinders. Eventually he made the switch, even developing a recording technique that produced records with considerably higher fidelity than those of his competitors. But consumers had already committed to a technology that was cheaper and good enough for their needs. This is a scenario that has played out time and again in the entertainment industries. The companies that have captured new markets often have done so by sensing opportunities early and moving in with "good enough" technologies that they have improved *after*, not before, locking consumers into their platform.

In the first two decades of the twentieth century, Victor and Columbia recognized that recordings, not the machines that played them, were their primary product. Adjusting accordingly, they positioned themselves in the middle of the market, which they recognized would give them maximum control and profit. On one side, they began hiring recording artists, which gave them upstream power over the music they were recording; on the other, they retained control of manufacturing, distribution, and promotion, which gave them downstream power over the sale of their music. Tin Pan Alley retained the job of managing copyrights for composers and lyricists.

Thanks to this strategy, recording royalties soon became a major moneymaker in the music industry. Victor alone saw record sales reach 18.6 million in 1915, and one estimate puts worldwide record sales earlier in that decade at about 50 million copies. In 1920, with World War I in the past, almost 150 million records were sold in the United States. The path ahead again seemed clear—until 1923, when broadcast radio emerged. Record sales then dipped for a few years, at one point threatening the survival of Columbia, but electric recording and playback emerged during the same period, and their vastly superior sound quality soon helped sales bounce back. By 1929, "gramophone fever" had struck, and the record business was booming.

Then came the Depression. Record sales in the United States took a precipitous dive between 1929 and 1933, falling from 150 million to 10 million. Sales of sheet music plummeted too, never to recover their importance in the revenue stream. To survive, companies merged, which brought about a wave of consolidations that transformed the industry into a genuine oligopoly. That transformation is described succinctly in a 2000 Harvard Business School case study titled "BMG Entertainment":

Edison went out of business. And the Radio Corporation of America (RCA), which had prospered as a result of radio's popularity, acquired Victor. In 1931, rivals Columbia, Parlophone and the Gramophone Company merged to become Electric and Musical Industries (EMI), based in England. The American operations of EMI passed into the hands of CBS, another radio network. The companies that emerged from the consolidation—RCA/Victor, EMI, and CBS Records—led the music industry in the following decades. Indeed, they formed the core of three of the five major music companies that dominated the industry in 1999.[3]

New recording companies, most notably Decca, emerged in the 1930s and the 1940s. But the industry remained under the tight control of a powerful few. "Between 1946 and 1952," the aforementioned case study reports, "the six largest companies produced 158 of the 163 records which achieved 'gold-record' status, and RCA/Victor and Decca represented 67 percent of Billboard's Top Pop Records chart."[4]

• • •

That kind of domination led to big profits—but also a vulnerability to the Next Big Thing, which arrived in the 1950s in the form of rock 'n' roll. At first the big companies simply didn't take the genre seriously. It seemed to be a fad, after all, and one that would appeal only to teenagers, a small niche audience with little money to spend. "It will pass," an expert on child development told the *New York Times* after describing what had happened with the Charleston and the jitterbug, "just as the other vogues did."[5]

Mainstream audiences were unimpressed by the quality of rock 'n' roll, too. "It does for music what a motorcycle club at full throttle does for a quiet Sunday afternoon,"[6] *Time* commented. Frank Sinatra felt even more strongly. "Rock 'n' roll smells phony and false," he told a Paris magazine. "It is sung, played and written for the most part by cretinous goons, and by means of its almost imbecilic reiteration, and sly, lewd, in plain fact, dirty lyrics ... it manages to be the martial music of every sideburned delinquent on the face of the earth."[7]

Sinatra wasn't alone in perceiving rock 'n' roll as immoral. Channeling questions being asked around the country, the *New York Times* asked: "What is this thing called rock 'n' roll? What is it that makes teen-agers—mostly children between the ages of 12 and 16—throw off their inhibitions as though at a revivalist meeting? What—who—is responsible for these sorties? And is this generation of teenagers going to hell?" The *Times* went on to give some credit for this to rock 'n' roll's "Negro" roots, which, it explained, provided the music with a "lustier" beat that had a "jungle-like persistence."[8] In the South, segregationists seized on this idea, one claiming that the music was a "Negro plot to subvert God-given values."[9] In the North, one prominent psychiatrist described it as a "cannibalistic and tribalistic" form of music and a "communicable disease."[10] Community leaders around the country called for boycotts of radio stations that played rock 'n' roll,[11] and government officials banned concerts, worried about the hysteria they brought on. "This sort of performance attracts the troublemakers and the irresponsible," Mayor John B. Hynes of Boston, declared. "They will not be permitted in Boston."[12]

Not everybody agreed. The disc jockey Alan Freed defended and promoted the music, arguing that it had a natural appeal to young people, who, he claimed, were better off in theaters, listening and dancing and letting off steam, than out on the streets making trouble. "I say that if kids have any interest in any kind of music," he told the *New York Times*, "thank God for it. Because if they have the interest, they can find themselves in it. And as they grow up, they broaden out and come to enjoy all kinds of music."[13]

Freed was more prescient than the big record companies, which worried that if they were to embrace rock 'n' roll they would alienate their main audience and tarnish their reputations. Given what they perceived as the music's niche appeal, inferior quality, and culturally threatening aura, they decided to stick with the cash cow they had been milking for years: the adult market.

That was a big miscalculation, of course. Rock 'n' roll took off. Small and nimble independent recording companies with little to lose stepped in. By 1962, forty-two different labels had records on the charts. The big companies finally woke up to their mistake and began playing catch-up by making big deals with rock 'n' roll performers (RCA signed Elvis Presley and Decca signed Buddy Holly), but their moment of blindness proved costly: in the second half of the 1950s, 101 of the 147 records that made it into the Top Ten came from independent companies. Temporarily, during the 1950s and the 1960s, the majors lost control.

But ultimately they won it back, because the economic structure of the music business favored concentration. Big companies were simply better equipped than small ones for long-term survival in the industry, which, as it grew in size and complexity, increasingly required an ability to leverage economies of scale in the market. Larger firms could more easily front the high fixed costs necessary to record music and promote artists, they could share overhead and pool risk across multiple projects, and they could leverage their size to exert bargaining power over promotional channels, distribution channels, and artists. Thus, as radio became an important means of promotion, the big companies had a distinct advantage. They had the power—and could arrange the payola—to guarantee that their music dominated the airwaves.

By the mid 1970s, the big companies had re-established themselves as the dominant force in the middle of the market, once again exerting upstream control over artists and downstream control over a diffuse network of relatively powerless distributors and promoters. During the 1980s and the 1990s, they gobbled up many of the smaller labels. In 1995, according to the Harvard Business School case study cited above, almost 85 percent of the global recording market was controlled by the six "majors": BMG Entertainment, EMI, Sony Music Entertainment, Warner Music Group, Polygram, and Universal Music Group.

As the 1990s came to a close, business was booming in all the creative industries. In music, records and tapes had given way to CDs, which turned out to be hugely profitable. How profitable? At the end of 1995, the International Federation of the Phonographic Industry reported that "annual sales of pre-recorded music reached an all time high, with sales of some 3.8 billion units, valued at almost US $40 billion." "Unit sales are currently 80 percent higher than a decade ago," the report continued, "and the real value of the world music market has more than doubled in the same period."[14]

• • •

For most of the twentieth century, the basic structure of the music industry remained the same. A group of businesses that had arisen specifically to produce and sell a single invention—the phonograph—somehow managed, over the course of several tumultuous decades, to dominate an industry that expanded to include all sorts of competing inventions and technological innovations: records of different sizes and qualities; high-quality radio, which made music widely available to consumers for the first time, and changed the nature of promotions; eight-track tapes, which made recorded music and playback machines much more portable; cassette tapes, which not only improved portability but also made unlicensed copying easy; MTV, which introduced a new channel of promotion and encouraged a different kind of consumption of music; and CDs, which replaced records and tapes with stunning rapidity. Through it all, with the exception of that one hiccup during the era of rock 'n' roll, the majors ruled. That's a remarkable feat. How did they pull it off? They used their scale to do two things very effectively: manage the cost and risk of bringing new content to market, and retain tight control over both the upstream and downstream ends of the supply chain.

Let's unpack this a little, starting with the management of risk. It is notoriously difficult to predict which people and which products will succeed in the creative industries. In a memoir, William Goldman summed up the problem when reflecting on the movie business: "Not one person in the entire motion picture field knows for a certainty what's going to work. Every time out it's a guess and, if you're lucky, an educated one." His conclusion? "Nobody knows anything."[15]

In practical terms, this meant, for much of the twentieth century, that in the hunt for talent, the creative industries relied on "gut feel." With little access to hard data about how well a new artist or a new album would do in the market, record companies, for example, could only make the most unscientific of predictions. They could put together focus groups, or study attendance figures at early concerts, but these were exceedingly rough measures based on tiny samples that were of questionable value when applied to the broader population. For the most part, the companies therefore had to rely on their A&R (artist & repertoire) departments, which were made up of people hired, optimistically, for their superior "instincts."

The big companies did agree on one element of success: the ability to pay big money to sign and promote new artists. In the 1990s, the majors spent roughly $300,000 to promote and market a typical new album[16]— money that couldn't be recouped if the album flopped. And those costs increased in the next two decades. According to a 2014 report from the International Federation of the Phonographic Industry, major labels were then spending between $500,000 to $2,000,000 to "break" newly signed artists. Only 10–20 percent of such artists cover the costs—and, of course, only a few attain stardom. But those few stars make everything else possible. As the IFPI report put it, "it is the revenue generated by the comparatively few successful projects that enable record labels to continue to carry the risk on the investment across their rosters."[17] In this respect, the majors in all the creative industries operated like venture capitalists. They made a series of risky investments, fully aware that most would fail but confident that some would result in big payoffs that would more than cover the companies' losses on less successful artists. And because of their size, they could ride out periods of bad luck that might put a smaller label out of business.

Scale also helped labels attract talent upstream. Major labels could reach into their deep pockets to poach talent from smaller ones. If artists working with independent labels began to attract attention, the majors would lure them away with fat contracts. All of this, in turn, set the majors up for more dominance. Having stars and rising talent on their rosters gave the major labels the cachet to attract new artists, and the revenue from established stars helped fund the big bets necessary to promote new talent.

For record companies, identifying and signing potentially successful artists was only the beginning of the job. Just as important were the downstream tasks of promotion and distribution. Once a company had invested in signing an artist and developing that artist as a star, it almost *had* to invest seriously in getting the artist's songs heard on the radio, making the artist's albums available in stores, and getting the artists booked as a warm-up act at big-name concerts. Record companies had to do everything they could to make people notice their artists, and their willingness to do that became an important way of attracting artists to their label. The majors didn't merely find artists; they used all the methods at their disposal to try to make those artists stars. Their decisions about promotion and distribution were gambles, of course, and the risks were big—which meant that, as on the upstream side of things, the "little guys" couldn't compete.

Consider the challenges of promoting a new song on radio. By the 1950s, radio had become one of the major channels available to record companies for the promotion of their music. But the marketplace was very crowded. By the 1990s, according to one estimate, the major record labels were releasing approximately 135 singles and 96 albums per week, but radio stations were adding only three or four new songs to their playlists per week.[18] Companies therefore had to resort to all sorts of tactics to get their songs on the air. Often this meant promising radio stations access to the major labels' established stars—in the form of concert tickets, backstage passes, and on-air interviews—in exchange for agreeing to play songs from labels' new artists.

The major record companies also practiced payola, the illegal practice of informally providing kickbacks to disk jockeys and stations that played certain songs. In the 1990s and the early 2000s, for example, the labels often paid independent promoters thousands of dollars to ensure, through a variety of creative promotional schemes, that the labels' new songs made it onto radio stations' playlists.[19] Typically, the major labels focused their efforts on the 200–300 stations around the country, known as "reporting stations," that sent their playlists weekly to Broadcast Data Systems, which then used the playlists to determine what records would make it onto the "charts."[20] In 2003, Michael Bracey, the co-founder and chairman of the Future of Music Coalition, memorably summed up the way things worked: "Getting your song on the radio more often is not about your local fan base or the quality of your music. It's about what resources you are able to muster to put the machinery in place that can get your song pushed through."[21]

Promotion is useless without distribution, however. For labels to make money, consumers have to be able to find and buy the music they have heard through promotional channels. And in the pre-Internet, pre-digital era, retail shelf space was very limited. Most neighborhood record stores carried only small inventories, perhaps no more than 3,000 or 5,000 albums. Even the largest of the superstores of the 1990s— glorious multistory spaces with whole soundproof rooms devoted to various genres—stocked only 5,000 to 15,000 albums.[22] As with radio, the majors therefore resorted to promising benefits in exchange for attention. To convince store managers to take a risk on devoting scarce shelf space to new music, the labels leveraged their stable of star artists, by offering access to in-store interviews, advance copies of albums, free merchandise, and more. And to make sure everybody noticed their "blockbuster" artists, the labels paid for special high-visibility placement for their releases in retail stores.

On the downstream side of the market, then, because of their size, power, and financial clout, the major labels were able to exercise tight control over both promotion and distribution. They owned the musicians and the music; they made the records, tapes, and CDs; and they dictated terms to radio stations and retail stores, which pretty much just had to go along. All of this, in turn, helped the majors maintain upstream control over artists, who had few options other than a big-label deal for getting their songs onto the necessary promotion and distribution channels, and who generally couldn't afford the costs, or incur the risks, of producing, manufacturing, and distributing their music on their own.

At the beginning of this chapter we asked why the same small set of companies has dominated the music industry for most of the twentieth century. The answer is twofold. First, the economic characteristics of the industry favored large firms that were able to incur the costs and risks of producing content, and were able to use their scale to maintain tight control over upstream artists and downstream processes for promotion and distribution. Second, until the very end of the century, no technological changes threatened the scale advantages enjoyed by the major labels.

Similar patterns emerged in the movie and book industries. At the end of the twentieth century, six major movie studios (Disney, Fox, NBC Universal, Paramount, Sony, and Warner Brothers) controlled more than 80 percent of the market for movies,[23] and six major publishing houses (Random House, Penguin, HarperCollins, Simon & Schuster, Hachette, and Macmillan) controlled almost half of the trade publishing market in the United States.[24] As with music, these publishers and studios controlled the scarce financial and technological resources necessary to create content ("People like watching shit blow up," one studio executive told us, "and blowing up shit costs a lot of money"), and they controlled the scarce promotion and distribution resources on the downstream side of the market. None of the technological advances that came in the twentieth century weakened these scale advantages.

By the 1990s this model was so firmly established in the creative industries, and so consistently profitable, that it seemed almost to be a law of nature—which is why, even two decades later, the executive who visited our class at Carnegie Mellon could so confidently declare that the Internet didn't pose a threat to his company's powerful place in the market. We think his confidence is misplaced, and in part II of the book we will explain why. But before we do so, we will take up something that should be understood first: the economic characteristics of creative content itself, and how these characteristics drive pricing and marketing strategies that are fundamental to the entertainment industries' business models.

3   For a Few Dollars More

When two hunters go after the same prey, they usually end up shooting each other in the back. And we don't want to shoot each other in the back.
Colonel Douglas Mortimer, *For a Few Dollars More*

Information Wants to Be Free   Information also wants to be expensive. Information wants to be free because it has become so cheap to distribute, copy, and recombine—too cheap to meter. It wants to be expensive because it can be immeasurably valuable to the recipient. That tension will not go away.
Stewart Brand, *The Media Lab: Inventing the Future at MIT* (Viking Penguin, 1987), p. 202

In chapter 2 we talked about the economic characteristics that drive market power in the creative industries. In this chapter we will talk about the economic characteristics of creative content itself, how these characteristics drive pricing and marketing strategies, and how these strategies might change in the presence of digital markets. We will begin by returning to 2009, when the head of market research at a major publishing house came to our research group with a simple but important business question: "What's an e-book?"

For years the publisher had followed the publishing industry's established strategy for selling its products. It would first release a book in a high-quality hardcover format at a high price, and then, nine to twelve months later, would release the same book in a lower-quality paperback format at a lower price. In the face of this established strategy, the publisher was saying: "I know where to release hardcovers, and I know where to release paperbacks. But what's an e-book, and where should I position it within my release strategy?"

Before coming to us, this publisher had released electronic versions of its books on the same date as the hardcover versions. However, it was questioning this decision, having seen several other publishers announce that they were delaying their e-book releases until well after the hardcover's release date in an effort to protect hardcover sales. For example, in September of 2009, Brian Murray, the CEO of Harper-Collins, announced that his company would delay the e-book version of Sarah Palin's memoir *Going Rogue* by five months after the hardcover release as a way of "maximizing velocity of the hardcover before Christmas."[1] Similarly, in November of 2009, Viacom/Scribner announced that it was delaying the release of Stephen King's new novel, *Under the Dome*, by six weeks after the hardcover release, because, as the company put it, "this publishing sequence gives us the opportunity to maximize hardcover sales."[2] Hachette Book Group and Simon & Schuster went even further, announcing in early 2010 that they would delay the e-book version of nearly all of their newly released "frontlist" titles by three or four months after the hardcover release.[3]

The implicit assumption each of these publishers made is that e-books are a close substitute for hardcover books, and that if an e-book is released alongside a hardcover edition many customers who previously would have purchased the high-priced hardcover will instead "trade down" to the lower-priced e-book.[4] This assumption seems reasonable on the surface; however, testing it is tricky, because we can only observe what actually happens in the market—we can't observe what would have happened if a book had been sold using a different strategy. Viacom, for example, could easily measure what hardcover and e-book sales actually were when they delayed the e-book release of *Under the Dome*, but it couldn't measure what those sales would have been if the e-book's release hadn't delayed (what economists refer to as the *counterfactual*). Much of the art of econometrics is in finding creative ways to estimate the counterfactual from the available data.

In the context of delayed e-book releases, one might try to estimate counterfactual sales by comparing the sales of books that publishers decided to release simultaneously in hardcover and e-book format against sales of books in instances in which the publisher released the e-book several weeks after the hardcover title. If e-book releases were delayed by different times for different books (for example, one week for some books, two weeks for some, and so on), a researcher could even run a simple regression, using the number of weeks an e-book was delayed (the independent variable) to predict the resulting sales of the hardcover edition (the dependent variable). That approach probably would work well as long as the undelayed books were essentially the same as the delayed books.

The problem is that delayed and undelayed books aren't the same. Publishers are more likely to delay e-book release dates for books they believe will sell well in hardcover format, which means the books that publishers release simultaneously in hardcover and e-book formats are fundamentally different in kind from books that they release in sequence. Thus, even if we observe a relationship between increased e-book delays and changes in hardcover sales, we don't know whether the changes in hardcover sales were caused by the delay of the e-book release or were merely correlated with differences in what types of books were delayed in the first place. Economists refer to this as "endogeneity"—a statistical problem that occurs whenever an unobserved factor (for example, the expected popularity of a book) affects both the independent variable (whether and how long books are delayed in e-book format) and the dependent variable (the resulting sales). Establishing a causal relationship in the presence of endogeneity requires finding a variable or an event that changes the independent variable without being influenced by the dependent variable.

The "gold standard" for establishing causation is a randomized experiment in which the researcher can vary the independent variable randomly and can measure the resulting changes in the dependent variable. For example, a publisher could randomly divide its titles into different groups, and then delay the e-book release of some groups by one week, some by two weeks, some by three, and so on. Unfortunately, these sorts of randomized experiments are extremely difficult to engineer for a host of reasons, among them the unsurprising fact that authors and agents object to having the work on which their livelihood depends become the subject of an experiment that may well cause their sales to drop. Indeed, we tried for several months to work with the publishing house mentioned at the beginning of this chapter to design a randomized experiment, but in the end we couldn't overcome concerns from the publisher's authors and agents about the how such an experiment might affect their sales.

If a randomized experiment isn't feasible, the next best option is a naturally occurring event that simulates the characteristics of the randomized experiment. And in 2010 just such an event happened. The publishing house we had been working with to design the experiment got into a pricing dispute with Amazon that culminated on April 1, when the publisher removed all its books from Amazon's Kindle store. Amazon was still able to sell the publisher's hardcover titles, just not the Kindle titles. Amazon and the publisher settled their differences fairly quickly, and on June 1 the publisher restored its e-books to Amazon's market and returned to its previous strategy of releasing its hardcovers and e-books simultaneously. Table 3.1 summarizes how the publisher's e-book releases were delayed during the dispute. A quick glance shows that the resulting e-book delays are close to the delays that might occur in a randomized experiment. Books that were released in hardcover format in the first week of the dispute (April 4) were delayed on Kindle format by eight weeks (June 1) as a result of the dispute. Likewise, books that were released in hardcover format the week of April 11 were delayed in Kindle format by seven weeks after the hardcover release, and so on for books released on April 18 (six weeks), April 25 (five weeks), all the way to books released the week of May 23 (one week). More important, neither the timing of the event nor the release schedule during the event was driven by the expected popularity of titles, and thus sales of undelayed books should provide a reliable measure of what sales of the delayed books would have been had they not been delayed. In this case, all we had to do to test how delaying e-books affected sales was to compare the sales of delayed books (those released

Table 3.1

Delays of Kindle releases of books during a major publisher's June 1, 2010 dispute with Amazon.

	Print release	Kindle release	Kindle delay (weeks)
Before April 1	Print and Kindle titles released same day		0
Week of April 4	April 4	June 1	8
Week of April 11	April 11	June 1	7
Week of April 18	April 18	June 1	6
Week of April 25	April 25	June 1	5
Week of May 2	May 2	June 1	4
Week of May 9	May 9	June 1	3
Week of May 16	May 16	June 1	2
Week of May 23	May 23	June 1	1
After June 1	Print and Kindle titles released same day		0

Source: Hailiang Chen, Yu Jeffrey Hu, and Michael D. Smith, The Impact of eBook Distribution on Print Sales: Analysis of a Natural Experiment, working paper, Carnegie Mellon University, 2016.

during the dispute) to the sales of undelayed titles (those released shortly before and after the dispute).[5]

But before discussing what we learned from analyzing the data, let's examine the economic rationale for why publishers have separate hardcover and paperback releases in the first place. Why make paperback customers wait nearly a year for their book? Why not release the hardcover and paperback versions at the same time? Why have two different versions at all?

At a high level, the answer to these questions is the default Economics 101 answer: Firms want to maximize their profit. But this goal is complicated by three economic characteristics of books and many other information-based products. First, the cost of developing and promoting the first copy of a book (what economists refer to as a product's fixed costs) is vastly greater than the cost of printing each additional copy (what economists refer to as a product's marginal costs).[6] Second, the value of a book can differ radically for different customers. Big fans are willing to pay a high price, casual fans are willing to pay much less, and many customers might not be willing to pay much at all. Third, consumers may not have a good idea of what they are willing to pay for a book in the first place. Books and other information goods are what economists refer to as "experience goods," which means that consumers must experience the product to know with certainty how valuable it is to them. This, of course, creates a problem for the seller. Once customers have read a book, they will probably be less willing to pay for it. Thus, the seller must strike a balance. On the one hand, the seller must provide enough information that consumers will know their value for the product; on the other, the seller must limit how much information is given to consumers, so that they will still want to purchase the product.

These characteristics cause sellers of books and other information goods to face several challenges in the marketplace. In this chapter, we'll focus on three specific challenges: extracting value for their products, helping consumers discover their products, and avoiding direct competition from closely related products.

Extracting Value

In a world in which customers have radically different values for a book and the marginal cost of printing an additional copy is very low, a publisher will extract the most profit from the market by convincing "high-value" customers to pay a high price for the book while still allowing "low-value" customers to pay a low price. But in a world in which customers are free to choose what product they buy, a company can't maximize profit if it can sell only a single product at a single price. If a publisher sells only at a high price, it will make money from high-value customers but will forgo income from low-value customers, who will buy only at a lower price. Alternatively, a publisher could generate income from both high-value and low-value customers by setting a lower price, but that would leave money on the table from high-value customers who would have been willing to pay more.

Of course, these statements aren't true only of books and other information goods; they apply in any market in which consumers have different values for a product. There are, however, two main ways in which these issues are more salient for information goods than for most other products. First, it is easier to vary the quality and usability of information goods than it is to vary the quality and usability of physical products. If you want to make a bigger engine, or a fancier stereo for a car, it costs money. But making a hardcover book costs only slightly more than making a paperback book. And for digital goods, the cost differences can be nearly zero. The cost of making a high-definition copy of a movie, for example, is nearly the same as the cost of making a standard-definition copy. Likewise, the cost of making a copy of a television show that can be streamed once is nearly the same as the cost of making a copy that can be downloaded and watched multiple times. Second, the fact that the marginal cost of producing additional copies of information goods is essentially zero opens up far more of the market than is possible for physical products. If it costs $15,000 to manufacture a car, anyone who isn't willing to pay even that marginal cost is excluded from the market. But if the marginal cost of producing an additional copy of a book is zero, everyone is a potential customer.

Thus, it is particularly important for sellers of information goods to find a way to maximize revenue from both high-value and low-value consumers. One way to do this is to convince consumers to reveal, either explicitly or implicitly, how much they are willing to pay—something that requires a set of strategies that economists call "price discrimination." To perfectly "discriminate" between consumers with different values for a product, publishers and other sellers of information goods would need to know exactly what each customer is willing to pay. With that information (and if they could prevent arbitrage between low-value and high-value customers), sellers could simply charge each customer their maximum price,[7] in the process extracting the maximum possible value from the market. The economist Arthur Pigou referred to this ideal scenario as "first-degree price discrimination."[8] Unfortunately for sellers, customers are rarely so forthcoming about their willingness to pay.[9]

Lacking perfect information about consumers' values, sellers are left with two imperfect options. First, a seller could set prices for different groups of consumers on the basis of an observable signal of each group's willingness to pay (a strategy economists refer to as "third-degree price discrimination"). For example, many operators of movie theaters discount ticket prices for students and senior citizens on the notion that these two groups of consumers generally have a lower ability to pay than other segments of the population and on the basis of the theater operators' ability to reliably identify these groups through age-based or membership-based ID cards.

Third-degree price-discrimination strategies are limited, however. Beyond age and some forms of group membership, there are few observable signals of willingness to pay that can (legally) be exploited, and for many products it is difficult to prevent low-value consumers from reselling their low-priced products to members of high-value groups.

In situations in which a seller can't use an observable signal of group membership to segment consumers, a seller can still adopt a strategy that economists refer to as "second-degree price discrimination." Here the seller's goal is to create versions of the product that are just different enough that a high-value consumer will voluntarily pay a high price for a product that is also being sold at a low price to low-value consumers. The hardcover-paperback strategy in book publishing is a classic example of second-degree price discrimination. Separate hardcover and paperback releases allow publishers to "discriminate" between high-value and low-value customers by relying on the fact that high-value customers generally are more willing to pay for quality (better binding and paper), usability (print size that is easier to read), and timeliness (reading the book as soon as possible after release) than low-value consumers. When this is true, releasing a hardcover book before the paperback version will cause high-value consumers to voluntarily pay a higher price for a book they know they will eventually be able to get for less money as a paperback.

The main concern when implementing this or any other second-degree price-discrimination strategy is to ensure that high-value customers aren't tempted to trade down to the lower-priced products. When is that temptation strong? When the quality of both products seems similar to consumers. This was exactly the concern of the publisher who approached us about its e-book strategy. Influenced by the conventional wisdom of some in the industry, it worried that consumers perceived e-books and hardcover books as similar products, and that releasing them simultaneously would reduce their valuable hardcover sales. What we found in the data, however, was that this conventional wisdom was wrong.

Our data from the natural experiment described above contained 83 control-group titles (titles that were released simultaneously in e-book and hardcover format during the period four weeks before and four weeks after the dispute) and 99 treatment-group titles (titles for which, during the dispute, the e-book was released one to eight weeks after the hardcover). The data showed that delaying the e-book release resulted in almost no change in hardcover sales for most titles. Most digital consumers, it seemed, did not consider print books a close substitute for digital ones—apparently, digital consumers were primarily interested in consuming a digital product. Put another way, consumers did not seem to perceive e-books as a lower quality version of hardcover books, but rather as a fundamentally different product.

Even more surprising was the effect on digital sales. Our data suggested not only that digital consumers weren't particularly interested in physical products, but also that they were quite fickle when the digital product wasn't available when they wanted to buy it. Counting all sales in the first twenty weeks after each e-book's release, e-book sales for delayed titles were 40 percent lower than sales of e-books released alongside the hardcover version. This suggests that e-books are a very different product from their print counterparts. It also suggests that when these digital consumers couldn't find the product they wanted to buy when they wanted to buy it, many of them simply left, and didn't come back. Maybe they just lost interest; maybe they found a different product to serve their needs; or maybe, even though they had initially been willing to pay for a legal copy, they sought out an easily available pirated one when the legal copy wasn't available. Whatever the case, our data suggested that the second-degree price-discrimination strategies that worked well for hardcovers and paperbacks didn't work as well for hardcovers and e-books.

Of course book publishers aren't the only sellers of information goods who rely on second-degree price-discrimination strategies. The music labels have a similar strategy. They release albums in both "regular" and "deluxe" editions. The extra content of the higher-priced "deluxe" edition is designed to attract fans who have a higher willingness to pay for this premium content, while still allowing the label to sell the regular album to lower-value consumers.

However, the second-degree price-discrimination strategies used in the publishing and music industries are nowhere near as complex as those used in the motion-picture industry. Figure 3.1 illustrates what was, until a few years ago, a typical release strategy for movies, with six main product-release windows, staggered over time and varying in quality, usability, and price. The first window, the theatrical window, is followed approximately 60 days later by hotel and in-flight airline services, then, after another 60 days, by a DVD release; then, six months to two years after the original theatrical release, the movie is made available through pay-per-view cable TV, pay cable networks, and advertising-supported television broadcasts.

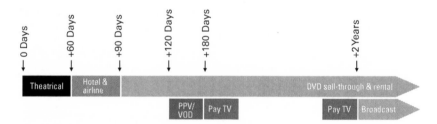

Figure 3.1
Typical release windows for movies in the period 1998–2005 (based on Industry sources and publicly available data).

Movie studios also use segmentation strategies within these release windows by varying usability (e.g., by offering separate versions for purchase and for rental) and quality (e.g., Blu-Ray versus DVD resolution, or bonus content versus regular content, similar to the "deluxe album" example above). For example, in 2005, when the movie *Lord of the Rings—Fellowship of the Ring* was released on DVD, New Line Cinema sold three DVD versions of the movie: a $30 two-disc widescreen edition for casual fans, a $40 four-disc "platinum series" special edition for serious fans, and an $80 collector's gift set for the most serious devotees.

As in book publishing, digital channels have added new complexities to the movie industry's established release windows (see figure 3.2 for a simplified version of current studio release schedules in physical and digital channels), raising many of the same questions for studios that publishers are asking: "How will digital sales channels such as iTunes affect sales in my other channels?" "How much do iTunes sales compete with iTunes rentals or with streaming services such as Netflix?" Most important, "How can I use these new channels to

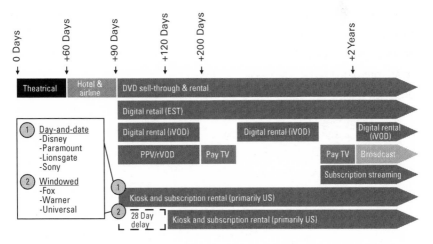

Figure 3.2
Typical release windows for movies in 2014 (based on industry sources and publicly available data).

enhance, as opposed to damage, my ability to differentiate between high-value and low-value consumers?"

As with publishers' e-book strategies, we believe the answers to these questions lie in the data—data that can help studios understand the interplay between product differentiation and sales cannibalization. However, these same data can also shed light on the other feature of differentiated-product strategies: the ability to create complementarities between channels so that sales in one window increase demand in subsequent windows.

Enabling Information Discovery

In early 2010, we observed how this kind of complementarity across release windows works while we were partnering with a major studio to help their decision-makers understand how broadcasting movies on pay-cable channels such as HBO, Cinemax, and Showtime affected DVD sales of these movies.

People at both the pay-cable channels and the studios felt that pay-cable broadcasts would substitute for DVD purchases and for purchases through other sales channels. Indeed, HBO was so concerned about the degree to which digital sales channels (notably iTunes) might cannibalize HBO viewership (and therefore HBO subscriptions) that whenever HBO licensed movies from studios it required the studio to remove those movies from all other sales channels (notably cable pay-per-view and iTunes) during the HBO broadcast window. DVDs were the exception to this rule, primarily because it was impractical to require retailers to take the DVDs off the shelves once they had been shipped to retail stores.

The fact that DVDs remained available during the HBO broadcast window gave us an opportunity to measure the effect of the HBO broadcast on demand for DVDs. To do this, we gathered weekly DVD and theatrical sales data over the lifetime of 314 movies broadcast on the four major US pay-cable channels (HBO, Showtime, Cinemax, and Starz) from January 2008 through June 2010.[10] Not surprisingly, "blockbuster" releases accounted for most of these sales. During the theatrical window, the top 10 percent of movies in our data captured 48 percent of all theatrical revenue, with the remaining 52 percent of revenue shared among the "obscure" titles—that is, all the movies in the bottom 90 percent. The data also showed that this popularity persisted nearly unchanged in the early DVD release window. The same movies that captured 48 percent of theatrical revenue also captured 48 percent of DVD revenue from the first month after release through the month before the movie was broadcast on pay-cable TV.

Why might a small number of movies dominate in the box office and early DVD release windows? It's possible that there are only a small number of truly good movies, and the concentrated sales figures simply reflect this; or that consumers move in packs, choosing to consume what they see their friends consume. However, we believe the concentration in movie sales is also influenced by how movies are released. Because movies are initially released exclusively in theaters, and because theaters have a limited number of screens to display movies, consumers are only able to discover a relatively small number of movies in the theatrical window. And because studios tend to promote DVD releases on the basis of theatrical performance, this skewed discovery continues in the early DVD release window—as seen in the sales data described above.

However, our data also showed a dramatic change after movies were shown in the pay-cable window. The pay-cable broadcast caused a movie's DVD sales to increase, but the increase was far larger among previously undiscovered movies—those in the so-called long tail. In the month after a pay-cable broadcast, sales of previously obscure titles increased to 65 percent of all sales (up from 52 percent in the preceding month).[11] What might explain this shift?

Our data show that pay-cable windows give consumers new opportunities to discover movies they hadn't discovered in the theatrical window. Specifically, our analysis showed that by the time most blockbusters (those in the top quartile of theatrical sales) entered the HBO window, 89 percent of their potential customers already knew about them, and hence there was little room for DVD sales to increase as a result of the pay-cable broadcast. Almost everyone who was going to be interested in the movie had already discovered it.

However, for the least popular movies—those in the bottom quartile of sales—the story was quite different. Only 57 percent of would-be customers had discovered those movies by the time the movie entered the pay-cable broadcast window. The remaining 43 percent of the movie's market had somehow missed out on discovering a movie that our data suggested they would enjoy. How did they miss out on these movies? One reason might be that these movies didn't have mass-market appeal and therefore weren't readily available in theaters. As we noted above, theaters can show only a certain number of movies at once, and to maximize their revenue will choose only movies that have broad market appeal. Because of this, some movies that will appeal to some customers are bound to slip through the cracks.

This may explain why customers didn't discover some movies *before* the pay-cable window, but it doesn't explain what changed *during* the pay-cable window. Why was the discovery process in the pay-cable window different than discovery in the preceding theatrical and DVD windows? One reason may be that, in contrast with the theatrical and DVD windows, in which you pay separately for every movie you watch, in the pay-cable window you pay no per-movie charge; once you pay the monthly subscription fee, you can watch anything on the network "for free." This ability to watch additional movies without an additional fee may allow pay-cable consumers to take a chance on movies they weren't willing to pay $15 to see in the theaters, allowing them to discover movies they wouldn't have discovered otherwise.[12]

But this sort of information discovery is valuable only if there is enough differentiation between the channels. For example, after consuming a product in the pay-cable channel, a customer may still want to buy it on DVD. If the products are too similar—for example, if consumers could easily record high-definition pay-cable broadcasts and watch the recordings whenever they wanted to—pay-cable broadcasts could compete with DVD sales instead of complementing DVD sales.

This brings us to our next marketing challenge for sellers of information goods.

Controlling Competition

In their book *Information Rules*, Carl Shapiro and Hal Varian use the example of CD-based telephone directories in the 1980s and the early 1990s to illustrate how competition can affect markets for information goods. In the mid 1980s, phone directories were controlled by the major phone companies and were licensed to high-value customers (such as the Federal Bureau of Investigation and the Internal Revenue Service) for about $10,000 a disk. However, as technology made it easier to digitize and duplicate information, these high prices attracted new competitors willing to invest the money necessary to manually copy the information in the phone companies' directories and make it available to the market. But once these competitors entered the market, the high-fixed-cost-and-low-marginal-cost economics of information goods took over, destroying the standard business models of generating huge profit from selling exclusive information to the highest bidders. Economic theory predicts that, in a perfectly competitive market for undifferentiated products, prices will fall to marginal cost. Not surprisingly, that is what happened in the phone-directory market. As new competitors entered, prices quickly fell to a few hundred dollars, and then to less than $20. Today, phone-directory information is, essentially, given away.

On the one hand, these lower prices for information are great for consumers, at least initially. On the other hand, marginal-cost pricing will hurt both producers and consumers if creators are unwilling to invest in new products for fear that they will be unable to recover their fixed-cost investments.[13] Indeed, the desire to encourage investment in markets for information-based products is the reason most modern economies give creators of information goods limited monopoly power over how their products are brought to market.

The creative industries, in turn, take this limited monopoly power and do exactly what economic theory says they should do: They use it to extract value from consumers by preventing direct competition, and by carefully controlling the quality, usability, and timeliness of how their products are made available. In the book business, the sooner a consumer can obtain a low-priced or free copy of a book, the harder it is for publishers to segment their markets on the basis of timeliness. In the music business, the easier it is for consumers to obtain the information found in bonus tracks and added features, the harder it is for labels to segment their markets on the basis of quality. And in the movie business, the easier it is for consumers to record and store content for future viewing, the harder it is for studios to segment their market on the basis of usability. This means that whenever someone asks "Why don't the creative industries make all their products available simultaneously in all potential distribution channels?" what that person really is asking is "Why don't the creative industries abandon all of their existing business models based on price discrimination and customer segmentation?"

The problem, of course, is that in many ways information technology has already dictated this choice. The very market characteristics that can weaken firms' control over customer segmentation—rapid information diffusion, easy information retrieval, and nearly costless information duplication and storage—are the basic capabilities of information-technology systems and networks, and their power is growing exponentially.

Today the critical challenge for people in the creative industries is determining how this technological change threatens their sources of market power and the profitability of their business models. That is the subject of our next chapter.

4 The Perfect Storm

Meteorologists see perfection in strange things, and the meshing of three completely independent weather systems to form a hundred-year event is one of them.
Sebastian Junger, *The Perfect Storm: A True Story of Men Against the Sea* (Norton, 1997)

Some of the guys get to where they feel invincible, but they don't realize that there's a real fine line between what they've seen and what it can get to.
Captain Albert Johnston, quoted in *The Perfect Storm*

Almost everyone knows the story—and the metaphor—made famous by Sebastian Junger's 1997 bestseller *The Perfect Storm*. The book recounted the saga of six veteran fishermen from Gloucester, Massachusetts, who decided, after weeks at sea in the fall of 1991, to sail home through dangerously stormy weather. Looking to protect a valuable catch, the men reasoned that they had survived many storms in the past and there was nothing special about this one. What they didn't recognize was that they were heading into seas buffeted by not just one storm but several, all of which had converged to create an unpredictable and monstrous weather event: a perfect storm. When the weather hit, the men responded with survival strategies that had always worked for them in the past but then discovered that they were battling conditions they didn't understand and weren't prepared to cope with. Overwhelmed, they perished at sea.

You know where we're going with this, right? For a long time, the creative industries—because they had evolved in market conditions that allowed a few big players to establish and maintain dominance—

had smooth sailing. When the occasional storm of technological change blew in, they knew how to ride it out—even how to use it to boost their own competitive advantage. But in the 1990s several very different kinds of change arrived all at once: a pervasive transition from analog to digital media, a boom in micro-computing and mobile technologies, and the advent of the Internet. The result was a new kind of turmoil that the creative industries simply hadn't evolved to cope with—a perfect storm of change that threatened their profitable business models and their established sources of market power.[1]

Change of this kind is difficult to foresee—particularly for incumbent firms, which tend to judge new innovations on the basis of the measures of success and profitability that apply to their existing business. Consider the tale that Howie Singer told us. Today, Singer is a senior vice president and the chief strategic technologist at the Warner Music Group, but in the 1990s, when that perfect storm of change hit the creative industries, he worked for AT&T. Singer and one of his colleagues, Larry Miller, sensed a great opportunity in that storm, and in 1997 they joined forces to co-found a2b Music, a service that made it possible to distribute compressed digital music files securely over the Internet. AT&T announced the launch as merely a trial, but Singer and Miller felt they were on to something big: a revolutionary new business that could transform how music was sold and consumed.

The a2b service was indeed new. To appreciate just how new, remember that the iTunes store was launched in 2003, the iPod in 2001, Napster in 1999, and the Diamond Rio (the first MP3 player) in 1998. The a2b team had a jump on them all. They were proposing that consumers anywhere with an Internet connection could download digital songs onto their computers, which they could then listen to wherever they wanted. In addition, the team had designed a portable music device to go along with the service that could play back an entire album's worth of music stored on a removable flash card—a remarkable technological feat at the time. To ease concerns in the industry about this fundamentally new means of distribution, the team had also devised a digital-rights-management protocol and had developed plans to (as they put it in announcing the service via press release) "incorporate micro-billing capabilities in future stages of a2b Music development and investigate other ways in which to integrate marketing strategies with retail outlets, so that the Internet can be used more efficiently and effectively for the emerging application of downloadable music."[2]

Singer and his colleagues thought they had done all that was necessary to help the music industry take charge of what was clearly its future: the Internet-based distribution of digital music. Enthusiastically, they began pitching their service to executives at the major music companies. They began by explaining that "a2b" referred to the shift from atoms to bits that was about to transform the music business. Thanks to coming advances in computing power and broadband connectivity, they continued, all music would soon be sold *only* as digital files, the management and sound quality of which would improve rapidly. The compact disc would become a thing of the past.

This ruffled some feathers. CDs were a source of great profits for the music industry in the 1990s, and their sales had been rising steadily for years. Why, executives asked, would we want to embrace a technology designed explicitly to replace our big moneymaker? One executive told the a2b team that he considered it an insult to hear his company's music referred to as "bits." Another—when presented with the idea that the a2b model would allow the labels to "disintermediate" and offer their music directly to consumers (which would have made it much harder for Apple to establish a beachhead with iTunes a few years later)—told the team they were speaking "a different fucking language."

Singer and his colleagues didn't have any more success winning hearts and minds when it came to demonstrating their technology. At a demo session for top executives at one of the major labels, they called up files of some of the company's most popular songs and played them through an expensive sound system. The songs had been encoded with an algorithm that produced smaller files and better sound quality than the MP3 standard that would later dominate the market.[3]

The team figured the executives would be wowed by how good the files still sounded, despite the compression, and by how much value consumers would get from being able to carry their music collections with them. But the executives didn't respond that way. Instead, they concentrated on the sound quality, which wasn't comparable to what one got from a CD. One executive, put off by what he was hearing, dismissed the whole effort with a line he must now deeply regret: "No one," he told Singer and his colleagues, "is going to listen to that shit."

Stymied by such reactions, a2b Music went nowhere, and AT&T ended its trial. Change was coming, of course. "It is feasible," Pekka Gronow and Ilpo Saunio wrote dreamily in *An International History of the Recording Industry* (1997), "that at some point in the future, recordings, as such, will no longer be produced at all, but that music will be supplied to the listener on request. ... In theory it should be possible to develop a sort of gigantic jukebox from which the listener could choose the music he wants at the time he wants it, working either by telephone lines, by cable, or on the air waves."[4] But at the time, this still seemed a distant prospect to most in the industry—the kind of thing for which no right-minded executive would consider abandoning the current cash cow.

A few years later, Apple stepped in; then came Rhapsody, Pandora, Spotify, and all sorts of other services. The rest is (still evolving) history. By turning the a2b team away, industry executives missed a chance to maintain their grip on one of the central pillars of their business model: the control of distribution.

"No one is going to listen to that shit." We didn't quote that line in order to make fun of the executive who uttered it. In his shoes, most people, whether they will admit it or not, would have reacted exactly the same way. As we have said, the effects of technological change are extraordinarily hard for market leaders to recognize—especially when, as in the above case, it represents a radical departure from how they established their market leadership in the first place. And even if you do see technological change coming, knowing what to do about it is still surprisingly difficult—as the publishers of the Encyclopaedia Britannica learned the hard way in the 1990s.

• • •

In 1990 the *Encyclopaedia Britannica* was riding high.[5] Over the course of more than two centuries, the company that owned it, Encyclopaedia Britannica Inc., had painstakingly built its reputation as the most comprehensive and authoritative reference work in existence. Sets of the encyclopedia sold for $1,500–$2,000 each and took up entire shelves in libraries and living rooms. They were luxury goods, but the company's enterprising in-home sales force had managed to convince droves of Americans that owning one was a requirement for—and a sign of—education, culture, and middle-class success. Production costs per set came to only about $250, and in 1990 the company made $650 million, its highest earnings ever. Britannica's prospects seemed bright. In 1989, staffers at Microsoft investigating the launch of a digital encyclopedia certainly thought so. In an internal strategy memo about the *Britannica*, they wrote "No other broad-appeal content product in any category in any medium has a well-established single-user price point anywhere close to this."[6]

Britannica sold not only encyclopedias but also an aura of trustworthiness and sober authority. Decades of research, planning, and editing went into the creation of each new edition of the encyclopedia (although a modestly revised "New Printing" was offered for sale each year in between editions, along with the highly profitable *Britannica Book of the Year*). Other companies produced encyclopedias that were smaller, cheaper, and friendlier to use, but the Britannica team didn't care. They wanted to serve customers willing to pay a premium for the very best product available—even if, as Britannica's own research revealed, most of those customers opened their encyclopedias at a rate lower than once per year. "These books aren't for reading," a sales manager once declared, "they are for selling."[7]

Much of Britannica's success in the decades leading up to 1990 had to do with the effectiveness of its sales team. Britannica representatives were hired carefully, were trained well, and, most important, believed in the value of what they were selling. Few teams did direct selling better, or with more sincerity, than the Britannica sales force (which had a strong vested interest in its product, with commissions of $500–$600 per sale).

Not surprisingly, when the personal computer arrived, in the early 1980s, Britannica's sales department dismissed it as a threat. In 1983, the department even produced a set of talking points on the subject for its sales representatives, who increasingly were encountering prospective customers who said they might prefer an electronic encyclopedia to a print edition. "One of the questions we are most frequently asked," the memo began, "by both our own people and outsiders, is, 'When will Britannica be available on a computer?' The answer we give is 'Not for a long time.'" The memo gave four reasons for that answer: Home computers didn't have enough capacity even to store just the index of the encyclopedia, much less the content. Storing the encyclopedia on a mainframe computer and then offering access via dial-up from a home computer would create an experience that was expensive, cumbersome, and slow. Only small portions of an article would be visible at any given time on a home computer's screen, creating a "disjointed" reading experience. Keyword searching, which would be critical for a digital encyclopedia, was a very blunt instrument.

Concerning that last reason, think of what would happen, the memo told representatives to ask consumers, if you searched for "orange" on a computerized encyclopedia. You'd get references to the color orange; the fruit orange; Orange County, California; William of Orange; and all sorts of other orange-related search detritus. You would then have to pick through that mess to figure out what might be useful to you—a laborious and time-consuming job that the editors of the print encyclopedia had already taken care of. "Britannica has already done all that work for you," the memo read. "Our indexers have read every article, analyzed what they read, and have determined exactly which entries should be in the Index. They have separated the colors from the fruits, etc., and have grouped the references accordingly. They have eliminated trivial references, so that when you follow an entry you can be sure you'll find a piece of relevant and significant information."[8] The print encyclopedia, in other words, was easier to use than any computerized version could be. Until that changed, the memo concluded, "we will not change our delivery method from the printed page to the electronic form."[9]

Whether Britannica liked it or not, however, the tide was turning. Two years later, in 1985, the company received an overture from Microsoft, which, after much study, had decided that a CD-ROM encyclopedia would be a "high-price, high-demand"[10] product that would diversify its product portfolio significantly. Microsoft proposed paying Britannica for non-exclusive rights to use its text as part of a multimedia digital CD-ROM of its own—and Britannica summarily turned the offer down. "The *Encyclopaedia Britannica* has no plans to be on a home computer," the company's director of public relations said at the time. "And since the market is so small—only 4 or 5 percent of households have computers—we would not want to hurt our traditional way of selling."[11]

"We would not want to hurt our traditional way of selling"—in those eleven words lay the seeds of Britannica's impending demise. But in 1985 that was a perfectly reasonable statement for the company to make. It had several strong incentives not to accept Microsoft's offer. For one thing, it feared the reaction of its sales representatives, who depended on the big commissions that selling print encyclopedias produced. If electronic copies of the encyclopedia were made available at a significantly lower cost, surely that would cannibalize sales of the print edition, which would lead to the departure of many highly trained sales representatives—one of Britannica's most important assets. The company also feared that a digital version of the encyclopedia would be perceived as a plaything and would undermine the aura of sober authority that Britannica had worked many years to establish. Home computers were a geeky novelty item at that point, too—hardly the sort of thing Britannica felt it should be risking its reputation on. In "The Crisis at Encyclopaedia Britannica," a 2009 Kellogg School case study of the Britannica's troubles, Shane Greenstein and Michelle Devereux cite yet another reason that accepting the offer didn't seem a good idea:

Britannica had no reason to take a risk with a young, unproven company like Microsoft, or to fear competition from it. After all, Britannica effectively controlled the top end of the encyclopedia market, charged the highest price premium among the encyclopedia publishers, and had strong and stable profits. The Britannica corporate culture was thriving, and the *Encyclopaedia* delivered strong returns. In fact, one former employee noted that "anyone who messed with the goose that laid the golden egg would have been shot."[12]

Britannica did recognize the potential value of an electronic encyclopedia, however—and so did its competitors. Grolier was first out of the gate, releasing a text-only edition of its encyclopedia in 1985, by which point Microsoft had committed itself to developing a multimedia CD-ROM encyclopedia. Not long after turning Microsoft away, Britannica began to develop its own multimedia CD-ROM encyclopedia—but not using its flagship product. Instead, it used text and images from a cheaper and less highly regarded encyclopedia that it also owned: *Compton's*, a reference work geared toward schools and schoolchildren.

The result, released in 1991 on disks for PCs and Macs, was *Compton's Multimedia Encyclopedia*. But what exactly was this product? Britannica didn't seem quite sure. It gave the disk away to customers who bought the print encyclopedia. That reassured its door-to-door representatives, because it suggested that the product was nothing more than a "sales closer"—a gimmick that wouldn't affect sales of the real encyclopedia. But Britannica also made the new *Compton's* available to the general public for $895, a price suggesting that this new encyclopedia in fact was a high-end competitor to the print encyclopedia. This dual approach failed on both fronts. Encyclopedia owners, led on by the sales force, assigned no value to *Compton's Multimedia Encyclopedia* and showed little interest in it, and general consumers judged $895 far too much to pay for an encyclopedia that was clearly second-rate. By 1993, after a series of price reductions that failed to elicit much interest, Britannica cut its losses and sold off the new *Compton's*, along with Britannica's whole NewMedia unit, to the *Chicago Tribune*, and decided to focus its efforts on an Internet-based version of the print encyclopedia, to be called Britannica Online. Meanwhile, however, sales of the print encyclopedia had begun to slide, going from $650 million in 1991 to $540 million in 1993. That was the year Microsoft launched Encarta, its CD-ROM encyclopedia.

Spurned by Britannica, and later by World Book, Microsoft had bought the rights to the moribund *Funk & Wagnalls New Encyclopedia*. From the perspective of the market for print encyclopedias, Funk & Wagnalls' text was significantly inferior to Britannica's, as was its reputation. However, in the new market, its text had an important advantage: It was composed in a consistent, almost modular format that lent itself to digitization, search, and hyperlinking much better than Britannica's voluminous text would have. This meant that Microsoft was able to produce a marketable CD-ROM quickly. Microsoft decided to scrap the name "Funk & Wagnalls" altogether, and instead focused on differentiating its new product by enhancing the text with graphics and sound, by investing in search technologies, by creating links that fed digital users' natural propensity for hopping from subject to subject, and by frequently adding and updating entries on current affairs. Microsoft didn't try to compete with Britannica on quality and reputation. Instead, it used the natural strengths of the new medium—video clips, search, hyperlinks, and frequent content updates—to stretch the idea of what an encyclopedia was, and to expand its potential audience. Encarta was to be a family product for use on home computers by parents and children alike, and it would cost only $99. The new approach worked. Microsoft sold 350,000 copies of Encarta in its first year, and a million the year after that.

Sensing danger, and with its print sales sliding, Britannica at last decided, in 1994, to create a CD-ROM of its flagship encyclopedia— only to meet with stiff resistance from its sales force, which once again argued that an electronic edition would cannibalize print sales. So the company revived the strategy it had deployed for *Compton's*. It made the digital *Britannica* available to owners of the print edition at no charge, and to the rest of the world for $1,200. As with the *Compton's* effort, this approach failed. Consumers balked at the price, and within two years Britannica was selling the CD-ROM for $200. Even then the company couldn't compete with Encarta, which offered a cheaper product that was more fun to use.

Encarta's content and reputation weren't nearly as good as those of the *Encyclopaedia Britannica*—not, at least, on the basis of the metrics established for the existing market for print encyclopedias—but for a lot of people it was good enough, particularly when combined with the new sources of value a digital encyclopedia could provide—graphics, sound, and search.

By 1996, sales of the *Encyclopaedia Britannica* had fallen to $325 million, half of what they had been only five years earlier. Not even Britannica Online—which, in a remarkably forward-looking and aggressively paced venture, had overcome a variety of technical obstacles and managed to put all 40 million words of the *Britannica* online—had been able to stop the decline. That year, with regret, Britannica's CEO, Joseph Esposito, sold the enterprise for $135 million to the Swiss financier Jacob Safra, who would prove unable to reverse the company's fortunes. In 2012, faced with the increasing popularity of Wikipedia (an encyclopedia whose content was generated by users, not by trained experts and professional editors), Britannica announced that it would no longer produce a print version of its encyclopedia. The *Britannica*'s run of more than 200 years had come to an end.

• • •

Why did Britannica have such difficulty responding to the changes in the market for encyclopedias? Britannica was, after all, the undisputed leader of the encyclopedia industry, with by far the most respected brand name, the most authoritative content, and the strongest sales force. How could its leadership in the market disappear after the entry of an unknown brand (Encarta) with inferior content (from Funk and Wagnalls) and without a commissioned sales force?

We believe the answer is that Britannica wasn't facing a single change to its business. Instead it was facing several factors that had the combined effect of radically changing the established sources of market dominance and the established models for selling content.

First, digital encyclopedias changed how value was delivered to consumers. Britannica's success and market power derived from its ability to deliver more value to consumers than its competitors could. This value came from high-quality authoritative text, careful editorial processes for approving content, pre-defined indexes to help consumers search content, and the social status communicated by having an expensive set of encyclopedia volumes in one's home. Digital encyclopedias didn't eliminate these sources of value, of course, but they substantially weakened them and introduced a new set of quality metrics: digital delivery; modular, easy-to-comprehend content; audio-visual material; rapid inclusion of new information; hyperlinking and digital searching; and social status that was increasingly communicated by having a computer, rather than a set of leather-bound books, in one's home.

The second factor that hurt Britannica was a fundamental change in how value was extracted from the market: a shift from a high-margin direct-sales model to a low-margin retail-sales model in which the content often could be bundled with, or even given away as a sales closer for, an entirely new product: the home computer.

The third factor was Britannica's market success in its established business: selling print volumes. Successful companies get successful by replicating and protecting their valuable business processes. For Britannica, this translated into a reverence for direct sales. Who ran the company and received promotions to positions of responsibility? Successful salespeople. Because of this, when a new way of selling content emerged, Britannica's leaders could only see it as a threat to their existing high-margin direct-sales strategy.

The fourth factor was a rapid shift in market power. It is important to remember that delay is usually not a bad thing for incumbent firms. Busy managers are faced with a steady stream of new business opportunities, and it's hard to fault them for not immediately adopting opportunities that are risky, unproven, lower in quality (at least from the perspective of how value has always been delivered in the market), and less profitable than the company's existing business. Recent research by Matt Marx, Joshua Gans, and David Hsu shows that, in most cases, incumbent firms are best served by taking a wait-and-see approach to new innovations: allowing the market to figure out which of the innovations is most likely to succeed, then either purchasing the innovator or partnering with it.[13] This is indeed an effective strategy in many circumstances. But it doesn't work if the entrant is able to quickly gain enough power in the new market so that the incumbent's assets are no longer valuable to a partnership—which is exactly what happened to Britannica. In an ironic coda to the story, when Esposito placed his company up for sale in 1996, he asked Microsoft (at the time a $60 billion company that employed the largest editorial staff in the encyclopedia industry) if it would like to make an offer for Britannica's assets. Microsoft declined.

What does all of this have to do with the entertainment industries? Possibly quite a bit. Later in the book, we will discuss how the entertainment industries are facing their own perfect storms of change. Technological change—in the form of long-tail markets, digital piracy, artists' increased control over content creation and distribution, the increased power of distributors, and the rise of data-driven marketing—presents the entertainment industries with a set of threats similar to those that Britannica faced. These threats include a new set of processes for delivering value to consumers, new business models for capturing this value, and difficult tradeoffs incumbent firms must make between protecting established businesses and exploiting new opportunities. And ultimately there is an even greater threat: new distributors that play increasingly active roles in the creation of entertainment content and that control customer attention and customer data, two increasingly important sources of market power.

None of these threats, by itself, would be likely to have much of an effect on the established structure of the entertainment industries. But together, we believe, they represent a perfect storm of change that is weakening the very sources of profitability and market power on which the entertainment industries have always relied, and is introducing new sources of profitability and power that the existing businesses and organizations are not well positioned to exploit. But just because we have used the perfect-storm analogy for the entertainment industries doesn't mean that we think they are doomed to the same fate as the Gloucester fishermen or the Britannica sales team. We are optimistic about the future of the entertainment industries—*if* they are willing to acknowledge and respond to the threats we have begun to discuss here.

But before discussing how to respond, we need to understand the nature of these threats in more detail. We'll start in the next chapter by discussing a new way in which companies are succeeding in the entertainment market: by using their customer connections and their data to develop a new set of processes for delivering value to consumers.

II Changes

Every time I thought I'd got it made / It seemed the taste was not so sweet
David Bowie, "Changes"

5 Blockbusters and the Long Tail

Very few entities in this world can afford to spend $200 million on a movie. That is our competitive advantage.

Alan Horn, chairman of Walt Disney Studios, quoted in Anita Elberse, *Blockbusters: Hit-Making, Risk-Taking, and the Big Business of Entertainment* (Holt, 2013)

It's easy to dismiss the random junk on YouTube as little threat to *The Sopranos*. ... But there is an audience for less-produced fare that can be made at a fraction of the cost of traditional TV programming.

Chris Anderson, *The Long Tail: Why the Future of Business Is Selling Less of More* (Hyperion, 2006)

The two recent management books quoted above—Chris Anderson's *The Long Tail* and Anita Elberse's *Blockbusters*—are often presented as opposite sides in the debate about how technology is changing the entertainment business. Anderson, a former editor of the magazine *Wired*, argues that the increased capacity of online sales channels (the so-called long tail) has shifted consumption away from markets dominated by a few "hit" products toward markets with many successful niches, and that firms in the entertainment industries should adapt their business models and marketing strategies to this new reality. Elberse, a professor at the Harvard Business School, has a decidedly different view. Drawing on case studies, market statistics, and interviews with executives in the entertainment industries, she shows that most of those industries' profits have always come from a small number of hugely popular titles, and contends that new technology is likely to increase, not diminish, the importance of "blockbuster" products to those industries.

We have great respect for the work of both Anderson and Elberse. But, as we'll show in this chapter, we believe they focused on the wrong question, at least in regard to how technological change is impacting market power in the entertainment industries. *Of course* long-tail products don't represent a threat to the "blockbuster" business model! By definition, long-tail products are products that very few people want to buy,[1] and it's hard to create a mass-market business whose goal is to create unpopular products. However, even if long-tail *products* don't pose a threat to the blockbuster business model, we believe that long-tail *processes* do.

That's what we want to explore in this chapter. We'll do that by focusing on technology's role in increasing the entertainment options available to consumers and then asking two important business questions: How do these new entertainment options create value for consumers? How can firms capture that value?

● ● ●

How do Internet markets create value for consumers? If you had asked this question in the late 1990s, the answer probably would have focused on the Internet's ability to reduce operating costs and increase market competition, leading to lower prices. In 1998 and 1999, we gathered data to test whether online prices were indeed lower than prices for the same products in brick-and-mortar stores. Working with Erik Brynjolfsson, we focused on a set of books and CDs sold by both brick-and-mortar and Internet retailers, collecting 8,500 price observations from 41 different retailers over 15 months. We found that online prices were between 9 percent and 16 percent lower than the prices charged by brick-and-mortar retailers—a significant source of economic value for consumers.[2]

Although our study design allowed us to compare prices in online and physical stores, it had a major limitation when it came to measuring the overall value consumers gained from online retailers. The Internet retailers in our study stocked nearly every book and every CD available, but brick-and-mortar booksellers generally stocked only the 40,000 to 100,000 most popular of the 2,300,000 books in print, and music stores only the 5,000 to 15,000 most popular of the nearly 250,000 CDs in print in 1999. Because the prices of products that aren't available can't be compared, we had to eliminate from our study all titles that weren't readily available in physical stores. Thus, although we were able to accurately measure the value online consumers gained from lower prices on relatively popular products, we were forced to ignore a potentially greater source of value provided by the Internet: the value consumers gained from being able to easily find and purchase the millions of books and CDs that were too obscure to be stocked in physical stores.

How much value do consumers gain from being able to access obscure titles online? Not much, many would argue. Consumers might be perfectly satisfied with the limited selection offered in brick-and-mortar stores. It is well known, after all, that small numbers of block-buster titles account for most sales of books, music, and movies in physical stores. Maybe this simply reflects a natural concentration of consumer tastes. Or maybe it reflects the economic characteristics of entertainment products, which, some have argued, naturally favor "superstars."

In their 1995 book *The Winner-Take-All Society*, Robert Frank and Philip Cook argue that many markets, including those for entertainment, have feedback loops that cause popular products to become more and more popular. Frank and Cook see three main factors driving this process: (1) People are naturally drawn to greater talent. (2) People like to consume the same content as their friends and peers. (3) Products with high fixed costs and low marginal costs are more profitable when sold in large quantities. William McPhee's 1963 book *Formal Theories of Mass Behavior* makes a similar point about the natural advantages of popular products, arguing that obscure products face a "double jeop-ardy" in markets and are thus likely to remain obscure: Most consumers aren't aware of their existence, and those consumers who are aware of them tend disproportionately to be experts who will also be aware of superior options.

On the other hand, even if a small number of blockbusters have dominated the entertainment markets in the past, that doesn't mean they will continue to do so. What seemed in the past to be natural market concentrations may actually have had more to do with the limitations of physical channels than with limitations in consumers' preferences. After all, you can't buy what you can't find. When consumers are offered a greater breadth of content, as they are on the Internet, what if it turns out that they have interests and appetites that are much more diverse than was previously assumed? From this perspective, the theories of McPhee and Frank and Cook both have significant limitations when applied broadly to entertainment markets. Consider the case of product differentiation. Economists recognize two main types of product differentiation: vertical and horizontal. In vertically differentiated markets, products exhibit a commonly agreed upon ordering of value. (Think of BMW versus Chevy, or Hilton versus Holiday Inn, or hardcover versus paperback.) For entertainment goods, one might argue that vertical differentiation exists in the context of James Joyce versus E. L. James, the Grateful Dead versus the Dead Milkmen, or Tom Hanks versus just about everyone else. But even here there is room for debate.

And that's the point. Many (perhaps most) entertainment goods don't have a commonly agreed upon ordering. This puts them in the category of horizontally differentiated products. Thus, while Frank and Cook's theory relies on consumers being drawn to "greater talent," and McPhee's relies on experts' awareness of "superior options," who can say whether your favorite book, movie, or song is "greater" or "lesser" than mine?[3]

Frank and Cook's second and third points are equally problematic when it comes to how technology might change the consumption of entertainment. Although people like to consume what their friends consume, online social networks allow us to receive recommendations from a wider circle of friends, potentially broadening our perspectives and allowing us to discover previously undiscovered niches. And although high fixed costs and low marginal costs naturally favor block-buster products, digital technologies can lower the fixed costs of production for many types of entertainment, reducing the scale necessary to profitably create some types of content.

So what should you do when the underlying theory is inconclusive? You should study the data. That's exactly what we did in 2000, with Erik Brynjolfsson and Yu Jeffrey Hu. And the answer that emerged was clear. Online access to niche products creates an enormous amount of value for consumers.

Our study began with an approach suggested by Madeline Schnapp, then the director of market research at O'Reilly Books. Previously, Schnapp had collected weekly sales data from Amazon for a set of O'Reilly's titles and matched these data to the sales ranks reported on Amazon's product page. With these data, she developed a model that, given knowledge of the sales rank of any of Amazon's titles, could predict the weekly sales of that title fairly accurately. Using a similar empirical approach and a dataset provided by an anonymous publisher, we replicated Schnapp's results and found strong evidence that online consumers had a great appetite for obscure titles. In our estimates, between a third and a half of Amazon's sales during that period came from titles that wouldn't be available in even the largest brick-and-mortar bookstores.

To calculate how much economic value was generated when consumers were given access to obscure titles, we turned to an approach for measuring the value of "new goods" developed by Jerry Hausman and Gregory Leonard. The main advantage of this approach is that it doesn't rely on theoretical views of consumer behavior or on judgments about the relative worth of mainstream and obscure titles. Instead, it focuses on the economic reality of what consumers purchase, and on their revealed willingness to pay for those purchases.

Adapting Hausman and Leonard's model to our setting, we found that the economic value consumers gained from being able to access obscure books online was between $700 million and $1 billion a year in

2000—nearly ten times the value consumers gained from lower online prices.[4] The main determinant of value for online consumers, in other words, wasn't saving a few dollars on products that they could already buy in physical stores. Rather, it was the value generated from their new ability to discover, evaluate, and consume millions of products that didn't fit within old brick-and-mortar business models.

This value increased during the 2000s, as we showed in 2008, when we used new data to revisit our earlier analysis. In the 2008 study, we found that three changes had significantly increased the value that consumers gained from the increased product variety they found online. First, Internet book sales had increased from 6 percent of total book sales in 2000 to nearly 30 percent in 2008. Second, in 2008 consumers were even more likely to buy niche titles than they had been in 2000. Third, consumers had many more books to choose from in 2008. The number of new titles printed each year had increased steadily from around 122,000 in 2000[5] to around 560,000 in 2008.[6] Our study showed that together these changes caused the value consumers received from increased product variety to quintuple between 2000 and 2008, to between $4 billion and $5 billion per year. And recent work by Luis Aguiar and Joel Waldfogel suggests that these figures may underestimate the true value gained from increased product variety online.[7] That's because, as we noted in chapter 2, no one knows which products will go on to become big hits. Publishers, labels, and studios do their best to anticipate which titles will fly off the shelves, but the process is imperfect. As a result, when technology gives previously undiscovered artists access to the market, some of these new artists—maybe even many of them—will surprise industry gatekeepers and land in the head of the sales distribution rather than the tail.

Aguiar and Waldfogel tested their theory by analyzing value creation from new music. They first observed that technological change has caused an explosion in new music, and that the number of new recorded-music products tripled from 2000 to 2010. Then, by applying their theory, they showed that the economic value created by these new products increases by a factor of 15 if one includes the possibility of blockbusters' emerging from the long tail.

One might ask whether these results extend to the tail of the tail—for example, to truly obscure books that have languished on the shelves of used-book stores for years. In some ways, these are exactly the sorts of products that McPhee's theory predicts will fail to deliver value. That's certainly Anita Elberse's view. In a 2008 article in the *Harvard Business Review* titled "Should You Invest in the Long Tail?" she writes (citing McPhee): "Although we might believe that 'the out-of-the-way book is at least a delight to those who find it,' in reality, the more obscure a title, the less likely it is to be appreciated."[8]

Does this effect show up in the data? A recent empirical paper by Glenn Ellison and Sara Fisher Ellison may shed some light on this question. Ellison and Ellison studied the economic value generated by online markets for used books, motivated in part by an experience one of the authors had had while searching for an obscure out-of-print title:

Several years ago, one of us wanted a thirty-year-old academic book on the pharmaceutical market which the MIT library did not have. The book had long been out of print, and looking for a used copy in brick and mortar stores would be like looking for a needle in a haystack. A quick search on Alibris, however, produced four or five copies for sale. A copy was ordered, for around $20, and it arrived shortly, with $0.75 written in pencil on the inside front cover and subsequently erased! The book had evidently been languishing on the shelf of some used bookstore for years, and not a single customer who noticed it was willing to pay even $0.75. A researcher needing the book happily paid $20 and would have paid significantly more.[9]

To test the possibility that similar things happen to other niche books, Ellison and Ellison collected a detailed dataset of the prices of used books from both online and physical stores. Their analysis of the data shows that the ability to find just the right title among the millions of obscure used books available online generates a great deal of economic value for both consumers and booksellers. In short, products which may not be appreciated by the vast majority of consumers, can still generate a great deal of delight—which economists equate to economic value—when discovered by the right consumer.[10]

• • •

If consumers derive an enormous amount of value from being able to find obscure products that match their tastes, as we have found that they do, that opens up many business opportunities for firms that can create these matches. But in order to capture that value, firms must first identify the specific business *processes* that are creating the economic value. What are the characteristics of information-technology-enabled markets that allow consumers to discover and enjoy products that weren't available in the scarce shelf space of brick-and-mortar retailers? To find out, we teamed up with Alejandro Zentner and Cuneyd Kaya who had obtained data from a major video-rental chain's physical and online stores. These data showed that rentals of the 100 most popular DVDs made up 85 percent of in-store transactions but only 35 percent of online transactions. But why? Is the online customers' shift toward obscure titles caused by the increased variety and ease of search offered online, or is it merely correlated with the types of consumers who choose to shop online rather than in physical stores? To answer that question, we needed to find an event that would cause consumers to shift from physical to online channels in a way that wasn't correlated with consumer preferences for obscure products. We found just such an event when our retailer began to close many of its local stores.

Because the decision about which stores to close wasn't driven by the preferences local consumers had for product variety,[11] we were able to isolate how an individual's consumption patterns changed when a local video-rental store closed and the person was forced to shift from the limited in-store selection to the expansive selection of the online channel. The data showed that giving consumers access to an expansive selection of products made them much less likely to rent blockbuster titles and much more likely to rent obscure titles that wouldn't have been available on the physical store's shelves.

We recognized, however, that this shift might be attributable either to supply (because consumers can access products that weren't available in physical stores) or to demand (because online search and discovery tools make it easier for consumers to discover new products). Separating these two effects requires fixing either the supply or the demand and varying the other—something we weren't able to do with our data. Fortunately, Brynjolfsson, Hu, and Simester managed to do just that in studying a different dataset.[12] They analyzed differences in the behaviors of online and catalog consumers of a women's clothing retailer that maintained the same product assortment in its online and catalog stores (thus fixing the supply side). They found that a significant part of the increased consumption of niche products comes from the demand side—that the technological characteristics of online markets can drive consumers toward niche products even when the supply side doesn't change.

Subsequent studies have examined in more detail the specific technological characteristics of online markets that might increase consumption of niche products. Consider the role of peer reviews in allowing consumers to evaluate obscure products. Some have argued that peer recommendations will result in more concentrated sales, because early tastemakers influence the market toward winner-take-all products. However, as we discussed above, peer recommendations could also allow consumers to discover new perspectives and ultimately buy more niche products, as Gal Oestreicher-Singer and Arun Sundararajan discovered when they collected and studied data from Amazon's product-recommendation networks. Their data allowed them to analyze the relative popularity of products in more than 200 categories of books on Amazon. They found that the categories that were more heavily influenced by peer recommendations exhibited much more diverse consumption patterns than other categories. Specifically, doubling the level of peer influence increased the relative revenue of the least popular 20 percent of products by about 50 percent and decreased the relative revenue for the most popular 20 percent of products by about 15 percent.[13]

Another factor that might shift consumption away from winner-take-all outcomes is the amount of product information available to consumers in online markets. When consumers have little independent information about products, they often follow the crowd and choose what other people are consuming. This behavior, which social scientists refer to as "herding," is well documented in the academic literature.

However, most studies of the sort described above have been conducted in artificial settings where consumers have very little outside information about the products they are evaluating. For that reason, we decided to investigate whether herding would persist in real-world markets in which consumers could easily collect outside information about the products they were evaluating. To do this, we partnered with a major cable company to conduct an experiment using their sales platform. We added a new menu to the company's video-on-demand service that displayed the most popular movies according to other consumers' recent ratings. In the default case, this menu displayed fifteen movies in descending order of the number of likes each movie had received from earlier users. However, at discrete times during the experiment we reversed the placements of two movies on the list. If user behavior is strongly influenced by the opinions of the herd, we would expect that users would rely on this incorrect information about a movie's "likes," and that the artificially promoted movies would remain in the artificially higher position or could even increase in popularity as they gained more followers and exposure.

We ran the experiment for six months in 2012, during which time more than 22,000 users purchased movies from our experimental menu of options. Our results showed little evidence of long-term herding behavior. When a movie was reported to have more or fewer "likes" than it really had, subsequent reviews by users caused it to return to its original position quickly. Moreover, better-known movies[14] returned to their original positions more rapidly than lesser-known movies.[15] In short, our experiment showed that consumers were less likely to follow the herd when they have access to outside information about the products they were evaluating—as might be expected among online consumers who can easily gather information about millions of different products.

Increased product variety, improved search tools, recommendation engines, peer reviews, and increased product information each play a part in driving online consumers toward niche products. But there is one other factor to consider: whether the anonymity offered by online transactions might change consumers' behavior by reducing their inhibitions.

Avi Goldfarb, Ryan McDevitt, Sampsa Samila, and Brian Silverman analyzed this effect in two contexts: purchasing alcohol and ordering pizzas.[16] (We know that alcohol and pizza aren't entertainment goods, but stick with us here.) The researchers found that when consumers purchased alcohol using a computer interface, they were more likely to choose products whose names were hard to pronounce than they were when ordering over a counter from a human clerk. Similarly, when ordering pizzas through a computer interface, consumers were more likely to order higher-calorie products and more complicated toppings than they were when ordering by telephone. Goldfarb et al. argue that the increase in online ordering of difficult-to-pronounce products probably occurs because of consumers' fears of "being misunderstood or appearing unsophisticated," and that inhibitions surrounding face-to-face orders for complicated, high-calorie pizzas probably are driven by consumers' concerns about "negative social judgment of their eating habits" and "negative social judgment associated with being difficult or unconventional."

What do alcohol and pizza have to do with consuming entertainment products? Beyond the obvious demand complementarities, it's easy to see how reduced social inhibitions in online transactions might also affect consumers' choices for entertainment. Katherine Rosman made this point in a 2012 *Wall Street Journal* article titled "Books Women Read When No One Can See the Cover," which documents the recent growth in demand for certain publishing genres. "Erotica used to be difficult to find," she wrote. "Chains and independent bookstores might have carried a few titles, but they were hidden away, and inventory was scarce." The anonymity offered by Kindles and other e-readers has changed all that. Think about the success of a long-tail product such as E. L. James' *50 Shades of Grey*.

Now, we understand why you may have choked on that last sentence. After all, *50 Shades of Grey* isn't a long-tail product. It has been translated into more than fifty languages, has sold more than 100 million copies, and has spawned a movie franchise. It's a classic blockbuster! You're right, of course—but you're also wrong. In many ways, *50 Shades of Grey* is a classic long-tail title. It was rejected by traditional publishing houses, it was brought to market not through a reputable print publisher but as a self-published e-book, and we wouldn't be talking about it today if it hadn't been aggressively promoted by passionate fans in online communities.

Here's the rub: *50 Shades of Grey*, like many other products today, has elements of both long tail and blockbuster. And in spanning those two categories, it highlights the limitations of focusing on products rather than processes, at least when it comes to understanding how technology is changing the entertainment industries.

• • •

In 2000, when we conducted our original research into how consumers gained value from online markets, our focus wasn't on the shape of the sales distribution, or on the proportion of sales in obscure products per se. Those measurements were means to an end: measuring the amount of value created by the online processes that allowed consumers to discover and purchase titles that weren't available on the shelves of brick-and-mortar stores.

But now, the discussion has shifted away from processes and toward the products themselves. Anderson's 2004 *Wired* article on the Long Tail spends a great deal of time documenting the proportion of products that sell at least one copy in a given month. Anita Elberse's 2008 *Harvard Business Review* article counters by showing that a large proportion of entertainment sales are concentrated in the most popular 10 percent or the most popular 1 percent of available products. A subsequent *Harvard Business Review* debate between Anderson and Elberse extends this discussion to whether long-tail products should be defined on the basis of the absolute number of titles stocked in brick and mortar stores or relative to the total number of titles available online.[17]

But, as we said at the outset of this chapter, processes, not products, are what we think the creative industries should focus on when evaluating the effects of long-tail markets. Does it really matter how flat the tail of the distribution is, or what proportion of sales resides in the flat part of the curve? No. Does it matter whether long-tail products are defined according to a relative or absolute measure of the stocking capacity of the market? Not really. What we believe matters is that consumers gain value from these long-tail products, and the processes necessary to capture this value differ from the processes the entertainment industries have relied on to capture value from blockbusters.

As we discussed in chapter 2, the entertainment industries' existing processes for capturing value from blockbusters start with a set of experts deciding which products are likely to succeed in the market. Once the experts have spoken, companies use their control of scarce promotion and distribution channels to push their products out to the mass market. In short, these processes rely on *curation* (the ability to select which products are brought to market) and *control* (over the scarce resources necessary to promote and distribute these products).

Long-tail business models use a very different set of processes to capture value. These processes—on display at Amazon and Netflix— rely on *selection* (building an integrated platform that allows consumers to access a wide variety of content) and *satisfaction* (using data, recommendation engines, and peer reviews to help customers sift through the wide selection to discover exactly the sort of products they want to consume when they want to consume them). They replace human curators with a set of technology-enabled processes that let consumers decide which products make it to the front of the line. They can do this because shelf space and promotion capacity are no longer scarce resources. The resources that are scarce in this model, and the resources that companies have to compete for, are fundamentally different resources: consumers' attention and knowledge of their preferences.

To be clear, we aren't arguing that long-tail products will replace blockbuster products. They won't. But we do believe that long-tail processes can and will be used to produce not only long-tail products but also blockbuster products. Netflix, for example, hasn't only enabled you to watch obscure movies that most of the world has forgotten; it has also produced *House of Cards*, *Orange Is the New Black*, and other hits of its own. This combination is extremely potent. Netflix—and other companies that effectively deploy similar processes—can capture consumer's attention by creating integrated digital platforms that offer a wide variety of content, can use proprietary data to predict what content will succeed in the market, and can take advantage of their unprecedentedly direct connections with consumers to promote this content directly to its likely audience. If you are a leader in the publishing industry, the music industry, or the motion-picture industry, the risk you face from the long tail isn't from products that don't sell well. The risk you face comes from the possibility that companies which specialize in long-tail products can adapt their processes—their platforms, data, and customer connections—to make it harder for you to capture value in the market for blockbuster products.

How might long-tail processes pose a threat to business models for blockbuster products? Consider the combined effect of the following technological shifts, which we will expand on in the next four chapters:

1. Digital piracy reduces the profitability of business models that use price discrimination to sell individual entertainment products, and also causes consumers to expect (and demand) the convenience of exploring many different products on a single site (for example, Netflix for streaming video, iTunes or Spotify for music, Amazon for books).

2. Technology gives previously disenfranchised artists new ways to reach their audience and new opportunities to create content, causing an explosion in the number of entertainment choices available to consumers.

3. Long-tail platforms develop sophisticated data-driven processes to learn consumers' preferences and to help them discover just the right content to meet their unique needs, generating significant consumer loyalty and market power for the platforms.

4. These data and processes become important resources in the entertainment business both for deciding which products will succeed in the market and for efficiently promoting this content to its audience, thereby giving the firms who control this data a significant competitive advantage over those that don't.

6 Raised on Robbery

You'll never stop [piracy]. So what you have to do is compete with it.
Steve Jobs[1]

We know piracy won't go away altogether, and we won't always agree on the best way to go about disrupting it. But we can agree on a vision for a digital future that better serves audiences and artists alike, and that future depends on reducing piracy.
Ruth Vitale and Tim League, "Here's How Piracy Hurts Indie Film," Indiewire[2]

In the 1980s, if you lived in a small town in India and wanted to see a movie, you had to wait two or three months after its release in a big city. When the movie arrived, you went to your local theater—a shoddy single-screen venue that might accommodate 1,000 people at a time. If the movie didn't come to your town, or if you just missed it, you were out of luck. All you could do was hope that it might be shown on television a few years later.

The emergence of the VCR in the mid 1980s changed everything. Videos proliferated, rental shops became commonplace, and "mini-theatres" with large VCR libraries began to offer screenings for a small fee. For Indian consumers, the change was glorious, even if the quality of the videos was third-rate. Now they could watch the latest films—or their favorite old ones—whenever they liked. Almost all of the videos in India were pirated, but what did that matter? This sort of casual piracy wasn't hurting anyone, was it?

• • •

Piracy is by no means a new problem in the West. (In the nineteenth century, the greatest source of pirated European books was the United States.) But as the creative industries developed during the twentieth century, the rich nations of the world developed and enforced an increasingly stringent set of copyright laws, in part to combat the threat of piracy. In large measure the system worked, at least in Europe and the United States, where most consumers were willing to obey the law and pay for the significantly better quality and easier availability of legal copies. But in poor and developing countries, the majority of the population simply didn't have the resources to find or buy these products legally, and piracy came to dominate the market.

The industries complained, of course. But as long as their copyrights were enforced and their profits remained strong in the developed countries, their executives slept well at night. Recordings, films, and books were physical objects, after all, and had to be reproduced one at a time. The process took time and cost money, and there were natural limits on the speed and quality of the pirated reproductions and on the ease with which these reproductions could be distributed. From the industries' perspective, piracy was illegal and annoying, but the products were hard to find and almost always inferior in quality, limiting any financial damage.

Everything changed in the 1990s with the "perfect storm" of technological change we discussed in chapter 4: the rapid growth of digital media, big advances in micro-computing and mobile technologies, and the advent of the Internet. Producing and distributing perfect copies of digital files suddenly became almost free, almost effortless, and almost ubiquitous. All those "natural" limits disappeared, and, almost overnight, piracy became an interconnected global phenomenon. The startling rise in 1999 of Napster, the peer-to-peer music-sharing site that allowed users all over the world to exchange music for free, presaged an ominous future for the creative industries. By some estimates, music revenue fell by 57 percent in the decade after the launch of Napster,[3] and DVD revenue fell by 43 percent in the five years after 2004, when BitTorrent gained popularity.[4]

The music industry, arguing that digital piracy represented a grave threat to its existence, mobilized a legal campaign to get Napster shut down, which succeeded in 2001. Building on this and other legal victories,[5] the creative industries banded together to convince American legislators to get involved. In 2011, Representative Lamar Smith of Texas introduced the Stop Online Piracy Act (SOPA), which proposed a set of restrictions and punishments that, he claimed, would help "stop the flow of revenue to rogue websites and [ensure] that the profits from American innovations go to American innovators."[6] The act failed because of surprisingly stiff resistance from technology companies and Internet activists, but the entertainment industries remained convinced that digital piracy was a major threat to their business that had to be met head on.

Many Internet activists and leaders of technology businesses disagreed. Sure, a lot of people around the world were now watching movies and listening to music for free, they said, but what evidence was there that this was hurting anyone? Perhaps the decreases in sales were attributable to changes in consumers' preferences and an increase in other entertainment options. And while the Recording Industry Association of American (RIAA) argued that 30 billion songs had been pirated in the years 2005–2010,[7] certainly the vast majority of the downloaders would have never purchased the music in the first place. Increased exposure due to piracy might even help artists by allowing new fans to discover their work. The popular press soon began echoing such messages. In 2013, summing up the findings of a widely cited study on piracy released by the Institute for Prospective Technological Studies,[8] CBC News wrote: "Entertainment industries are beginning to realize that the sharing of films and music online generates marketing benefits and sales boosts that often offset the losses in revenue from illegal sharing of content."[9]

Some early academic studies reinforced these points. Early theoretical models showed that piracy could benefit the industry by removing the most price-sensitive consumers from the market, by establishing an initial customer base, and by increasing overall diffusion and product awareness in the market.[10] One of the earliest empirical papers on the subject, published in the highly respected *Journal of Political Economy*, found that music piracy had no effect on legal sales.[11]

So maybe piracy actually isn't such a bad thing. Indeed, what if it is actually *increasing* overall consumption by generating unprecedented amounts of buzz, attracting new audiences to concerts and merchandise, and encouraging at least some portion of a vastly expanded global audience to buy legal copies of what they had just consumed for free? And what if, while serving a useful function as a discovery mechanism, it is also forcing the creative industries to lower their prices and increase the availability of their products? Shouldn't that boost sales and benefit society? Isn't that exactly what happened with iTunes and digital downloads? Moreover, haven't the creative industries always complained that product sharing based on new technologies is about to destroy their profitability—and haven't they always been wrong? Why should this moment be any different?

All those questions really are variations on the same question: Does piracy cause harm? In what follows, we'll explore that question as it affects first producers and then consumers.

Does piracy harm producers?

At first, answering this question seems simple. If consumers can use pirated content to get their music and movies for free, *of course* they will purchase less content as a result. Isn't that why music sales fell so dramatically after Napster was released? Consider the graph reproduced here as figure 6.1, produced by Alejandro Zentner in 2006, which illustrates the rise and fall of global music sales between 1990 and 2003.

Something clearly happened around 1999, when sales began a four-year decline that reduced total revenue by almost 25 percent (from a peak of $40 billion to about $31 billion). Digital piracy is a reasonable suspect here, in view of the correlation between the decline in sales and Napster's rise to prominence in 1999.

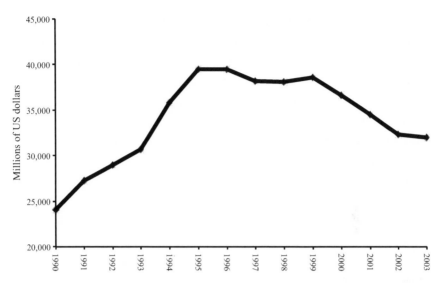

Figure 6.1
The rise and fall of global music sales between 1990 and 2003.Source: Alejandro
Zentner, "Measuring the Effect of File Sharing on Music Purchases," *Journal of Law
and Economics* 49, no. 1 (2006): 63–90.

However, a correlation between the rise of Napster and falling revenue in the music industry doesn't mean that digital piracy *caused* revenue to fall. Revenue can rise and fall for all sorts of reasons, and there
are plenty of other potential explanations around the turn of the century that should be considered. Between 1999 and 2003, for example,
the expansion of broadband Internet access created a wealth of new
entertainment possibilities for consumers, who for the first time could
easily while away hours online browsing the Web, playing games, chat-
ting with friends and strangers, spending time on social networks, and
more—things that cut into the limited time that they had available for
music, books, and movies. Maybe what caused the decrease in sales
shown in figure 6.1 was simply a shift in how consumers were spending
their leisure time, not a shift in how they were acquiring their music.
Or maybe the decline in sales was just a natural by-product of the shift
from album sales to unbundled digital-single sales. And who can say
that consumers would have made more purchases if piracy hadn't been

an option? Students, after all, typically do a lot of pirating (because they have the time and the technical savvy), but, with their limited income, maybe they wouldn't have purchased the content they were pirating. And if that's the case, so the argument goes, why not just let them be pirates, and assume that by freely sharing files they are helping media companies run a global social-marketing campaign?

If this has been a frustrating discussion, that's because it has addressed a frustrating question. A strong theoretical case can be made that piracy will cause legal sales to decline, but a strong theoretical case can also be made that piracy will have no effect or even increase sales. And using data to analyze the effect of piracy on sales is tricky. It's easy to observe the decline in music sales after Napster, but establishing that Napster caused sales to decline requires a reliable estimate of what sales would have been in the absence of piracy—a counterfactual estimate of the sort we discussed in chapter 3. Randomized experiments would be ideal for such an exercise, but they are nearly impossible to conduct—it's hard to imagine bringing piracy to a temporary halt for a random set of products or among a random set of customers.

Lacking a randomized experiment, one might simply make comparisons across products or across consumers: Do heavily pirated products have fewer sales than other products? Do consumers who pirate heavily purchase fewer products than other consumers? These approaches, unfortunately, suffer from the same endogeneity problem we discussed in chapter 3. When it comes to sales, heavily pirated products are systematically different from less pirated ones, and the sorts of consumers who pirate a lot are systematically different from consumers who don't pirate. Because of this, the observed sales of less pirated products don't make a good counterfactual estimate for what sales of heavily pirated products would have been if piracy hadn't been available, and likewise for consumers.

Despite these obstacles, researchers have managed to design and conduct studies that can help us determine the effect of piracy on legal sales. Many of these studies use uncorrelated events, known to econometricians as "instruments," to simulate the effect of a randomized experiment. For this to work, the instrument has to significantly change the ease of pirating and has to do so in a way that isn't directly correlated with legal sales. For example, in the *Journal of Political Economy* paper mentioned above, Felix Oberholzer-Gee and Koleman Strumpf used German school vacations to measure the effect of piracy on music sales in the United States in 2002. Why use a German school vacation to study piracy in the United States? Because, they reasoned, when German students are on vacation they have more time for pirating and can share music files more easily with people in the United States, and because German school vacations are otherwise uncorrelated with music sales patterns in the United States. Other researchers have used different instruments (the rollout of broadband Internet access across cities, the availability of a network's content on iTunes, or the imposition of anti-piracy regulations in a particular country) to gauge piracy in a way that isn't correlated with sales.

None of the aforementioned studies is perfect, and all empirical studies are limited by the statistical methods used, by what data are available, and by how well those data generalize to other settings. Because of this, the best way to get a sense of the academic literature is to take the broadest possible look at the published results and see how often the same result shows up in different contexts. That's exactly what we and Brett Danaher attempted to do in two publications: a chapter we contributed to *Innovation Policy and the Economy* (an edited volume published in 2014 by the National Bureau of Economic Research)[12] and a paper we presented to the World Intellectual Property Organization in November 2015.[13] In these papers, we surveyed all the peer-reviewed journal articles we could find that had studied the effect of piracy on sales. We found 25 such articles.[14] Three of them reported instances in which piracy hadn't affected sales; 22 reported instances in which piracy had harmed legal sales significantly. (The 25 papers are listed in a table in the appendix to this chapter.)

For a complicated question such as the effect of piracy on sales, 22 out of 25 represents a remarkably powerful consensus in academia.[15]

For all intents and purposes, among scholars the matter is settled. In the vast majority of cases, piracy has exactly the effect you would expect it to have. It reduces paid consumption by allowing consumers who otherwise would have purchased the content to get it for free.

And the effect of piracy on sales tells only part of the story. The problem, at least from the perspective of the entertainment industries, is that piracy not only reduces sales but also makes it harder for these industries to extract revenue from their remaining consumers. That's because piracy creates a new alternative for consumers that competes on not just price but also on timeliness, quality, and usability—the same factors the entertainment industries have relied on to execute their business models. As we discussed in chapter 3, the ability to control when and how products are released is critical to how the entertainment industries make money. With piracy as an option, consumers who normally would have to wait several months after a movie leaves the theater before they can buy it on iTunes (for 10–15 dollars in standard definition and 15–20 dollars in high definition), and several more weeks for the opportunity to rent it (for 3–5 dollars), can now obtain a pirated copy of the movie that is free, is in high-definition format, can be watched on almost any device, and is typically available immediately after—or in some cases a week or two *before*—the legal version. This, in turn, forces producers to lower their prices and to change their release strategies.

Of course, from a consumer's perspective this looks like a great thing. No longer do they have to wait months to get the content they want in the format they want it. No longer do they have to pay high prices for something that costs almost nothing to reproduce. Even though piracy is bad for producers, it's certainly good for consumers. Isn't it?

Does piracy harm consumers?

Well, it's complicated—in many ways more complicated than the question of whether piracy harms sales. Some scholars argue that even if piracy harms sales, the lost sales simply represent a transfer of wealth from producers to consumers, and that focusing solely on lost sales ignores the potential benefits that piracy can provide—notably benefits that come from allowing consumers to access content they wouldn't have purchased in the first place. Indeed, it's easy to show that if piracy doesn't change the production of new content, then consumers gain more from piracy than producers lose.[16]

But that's a big "if." What if, by reducing the revenue producers were able to make from selling content, piracy caused some types of content to no longer be profitable? Wouldn't that harm consumers? The International Federation of the Phonographic Industry has made this very case, pointing out that music is "an investment-intensive business" and arguing that piracy "makes it more difficult for the whole industry to sustain that regular investment in breaking talent."[17] Certainly if labels, studios, and publishers know that piracy will reduce the amount of money they can make on some types of investments, they will be less willing to invest in the first place. And if the entertainment industries reduce their investment in new content, that will hurt consumers in the long run.

Intuitively, this argument makes pretty good sense. But it turns out to be very hard to measure, for several reasons. First, it's hard to isolate the effect of piracy on investment, because the same technological advances that facilitate piracy also reduce the cost of production and open new creative options for individual artists, both of which may increase overall investment and increase the industry's output. It is also difficult to develop a reliable measure of "investment" in entertainment. In some industries—for example, pharmaceuticals and biotechnology—you can measure innovation by studying the number of patents issued each year. The creative industries are different. Generally, you can measure innovation only very approximately, by studying the number of books, movies, or albums released. But volume alone isn't a particularly useful measure in these industries, in view of the increasing importance of the long tail—which itself poses measurement problems. One option, therefore, is try to measure changes in the production of the popular and high-quality content the entertainment industries have always produced.

Unfortunately, those changes also are hard to measure, because measuring them requires adjusting the volume of production to a specific level of quality—something that is inherently difficult. But in 2012, Joel Waldfogel addressed that difficulty directly in a study of how Internet piracy affected the supply, the creation, and the quality of music releases after 1999, the year Napster was launched.[18] Waldfogel's study is instructive, so let's unpack it a little here.

To measure quality—admittedly, a subjective measure—Waldfogel relied on the wisdom of the crowd. Specifically, he turned to the assessments of professional music critics as represented in 88 different retrospective "best of" rankings (e.g., *Rolling Stone*'s 500 best albums). The index covered the period from 1960 to 2007; drew from lists that appeared in the United States, England, Canada, and Ireland; and included more than 16,000 musical works. Figure 6.2 shows the pattern that the index revealed: a rise in quality between 1960 and 1970, a fall between 1970 and 1980, a rise in the mid 1990s, a fall in the

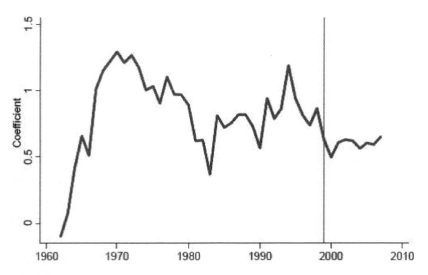

Figure 6.2
A critic-based quality index.Source: Joel Waldfogel, "Copyright Protection, Technological Change, and the Quality of New Products: Evidence from Recorded Music since Napster," *Journal of Law and Economics* 55 (2012), no. 4: 715–740, figure 3, page 722.

latter half of the 1990s, and a leveling off after 2000. Because his index showed that quality was already falling well before the emergence of Napster in 1999, and because quality stabilized soon thereafter, Waldfogel concluded that his data provided "no evidence of a reduction in the quality of music released since Napster."[19] He created similar indices of quality using radio play and sales, and each confirmed his conclusion.

What might explain this result? Why did the quality of music produced remain steady while industry revenue declined sharply? Shouldn't less money lead to less investment and to lower-quality music?

One explanation is that the rise of music piracy wasn't the only force affecting the industry from 1999 to 2008. A technological revolution took place at the end of the 1990s—one that changed the fundamentals of the music business by dramatically reducing the costs of creating, promoting, and distributing music. Today artists can use off-the-shelf software to produce recordings that rival in quality the recordings that in the past only expensive studios could produce. In addition to production, technology has also democratized promotion, with sites such as Pandora and Last.fm providing low-cost promotion to all sorts of artists. The same goes for distribution. In the days of CDs, as we noted in chapter 2, a few large labels and retailers dominated the market, and artists had to rely on them to distribute their music. But today artists can distribute their music—on iTunes, for example—for almost nothing.

In short, studying changes in the quality of new music after Napster was released can only tell us what happens when piracy becomes available at the same time as other technological innovations that lower the costs of producing, promoting, and distributing content. We don't know whether piracy would have hurt the supply of content had the costs of production, promotion, and distribution not changed. To try to answer that question, let's take a closer look at the movie industry. Movies cost far more to make than songs, so one would expect that a sharp decline in revenue would lead to a larger decline in production and supply for movies than would occur for music. This should make it easier to identify the effect of piracy, but it leaves us with the same problem we had when studying the effect of piracy on music sales after 1999. The same business-expansion and technology shocks that rocked the music industry also rocked the motion-picture industry, and these shocks coincided with the rise of piracy.

What we really need is to study a setting in which a new technology enabled piracy but didn't significantly change the other aspects of the business. For that we have to go back to the mid 1980s, when the VCR arrived in India.

India has been one of the world's major producers of movies since the early 1900s. Profits are a major motivation in the movie industry, and producers have long entered and exited the market freely. Before the VCR, movie piracy was very difficult and almost non-existent, and thus we should be able to observe the effects of VCR-based piracy on both demand and supply in the 1980s and the early 1990s—especially insofar as no technological change during that period significantly changed the cost structure of movie production and distribution in India.

What do the data from this period tell us? In 2014, we worked with Joel Waldfogel to compare data on the Indian film industry from before 1985 against data from the period 1985–2000. That study began by showing that industry revenue declined sharply with the advent of VCR-based piracy—no surprise, of course, in view of the consensus among researchers that piracy almost always harms sales. However, our paper also analyzed how the Indian film industry's production changed during that time period, and what we found will surprise those who view piracy as a victimless crime. The data showed that after 1985 there was a significant decrease in the number of movies made in India (figure 6.3) and a significant drop in their quality as measured by IMDb ratings (figure 6.4).[20]

The best explanation for both of these types of decline, we concluded, was the advent of VCR-based piracy. This explanation, in turn, led to an important new finding: that—at least in India between

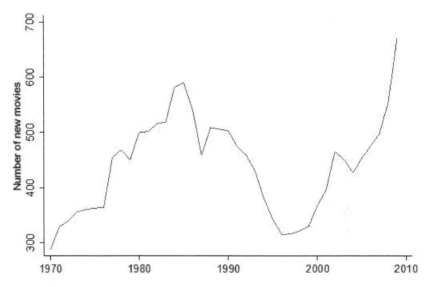

Figure 6.3
India's annual production of movies.Source: IMDb, 1970–2010.

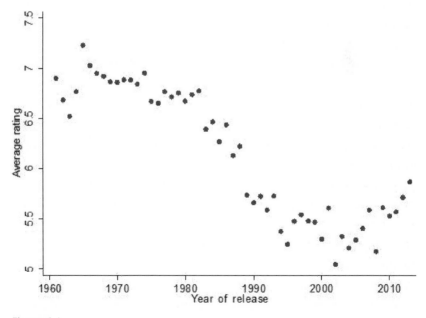

Figure 6.4
Users' average ratings of Indian movies on IMDb.

1985 and 2000—piracy did indeed reduce the incentive to create new content.

• • •

This brings us back to where we started this chapter. For someone growing up in India in the 1980s, or even for someone growing up in a suburb of Washington, it was easy to imagine that casual piracy wasn't hurting anyone. But years later, after carefully analyzing the data, we have come to the opposite conclusion: Not only does piracy hurt producers; it can also hurt consumers if some types of content are no longer profitable for producers to make.

What's the solution? That depends on what you mean by "solution." If you are looking for a solution that eliminates all digital piracy, you are out of luck. As long as content can be digitized, people will find ways to make copies and share these copies with friends and strangers online. And because there is no perfect solution to stopping digital piracy, some might argue that policy makers should give up trying and instead should figure out how to help redistribute revenue from one industry to another. Nick Bilton, a technology correspondent for the *New York Times*, advanced that argument in 2012.[21] In an op-ed piece titled "Internet Pirates Will Always Win," Bilton made it clear that he thinks trying to fight piracy is pointless. "Stopping online piracy is like playing the world's largest game of *Whac-A-Mole*," he wrote. "Hit one, countless others appear. Quickly. And the mallet is heavy and slow." That's an appealing metaphor, and it has some merit. Content-identification systems can be fooled, shutting down some sites only causes new sites to open, and new file-sharing protocols do make it harder to monitor piracy. But in laying out his argument Bilton didn't take into account the history of online price competition. Think back to 1998, when the conventional wisdom was that the Internet would allow consumers to easily find the lowest price online—an argument that Robert Kuttner summed up in the May 11, 1998 issue of *Business Week*: "The Internet is a nearly perfect market because information is instantaneous and buyers can compare the offerings of sellers worldwide. The result is fierce price competition, dwindling product differentiation, and vanishing brand loyalty."[22]

Kuttner's argument seemed plausible enough at the time. Why would you pay more for something that you could easily find for less? But the argument ignored something important: product differentiation. If you can differentiate your product on the level of reliability, convenience, service, quality, or timeliness, consumers will often cheerfully pay more for a product they know they can get elsewhere for less.

For example, in joint work with Erik Brynjolfsson, we analyzed consumers' behavior at Internet price-comparison sites, also known as "shopbots."[23] Our data allowed us to study how shopbot users, arguably among the most price-sensitive consumers online, behaved when shown a set of competing offers. The data revealed that consumers were willing to pay several dollars more to purchase a book from Amazon even when lower-price alternatives from lesser-known places like 1bookstreet, altbookstore, and booksnow were displayed on the same search page, just a click away.

What does this have to do with anti-piracy regulation? Possibly, quite a lot. Think about the digital-media space. Then imagine that the creative industries and their legal online distribution partners play the role of Amazon in the example above, and that the pirate sites play the role of lower-price alternatives such as 1bookstreet, altbookstore, and booksnow. With the proper differentiation—say, if a producer uses iTunes and Hulu to improve the convenience, quality, and reliability of how it distributes its products—the creative industries should be able to convince some people to pay for their content through legal channels even if those people know that free pirated alternatives are available.

But "should be" doesn't count as evidence. Is there empirical evidence that making content available on sites such as iTunes and Hulu causes some consumers to switch from piracy to legal consumption? We studied this question in two settings and found that the answer is Yes.

In the case of iTunes distribution, we worked with a major motion-picture studio to analyze how piracy on its older, "catalog" movies changed when they were added to iTunes. Our data included more than 1,000 catalog titles released on iTunes in 48 countries from February 2011 to May 2012. Our results showed that releasing these movies on iTunes caused demand for pirated copies to decrease by 6.3 percent relative to demand for pirated copies of similar control-group movies that didn't experience a change.

We found similar results for Hulu streaming availability by studying how piracy changed after ABC added its television content to Hulu (as it did on July 6, 2009). In our analysis, we compared piracy levels for the nine series that ABC added to Hulu against piracy levels for a control group of 62 series that experienced no change in Hulu availability. The results are summarized in figure 6.5, which compares piracy levels

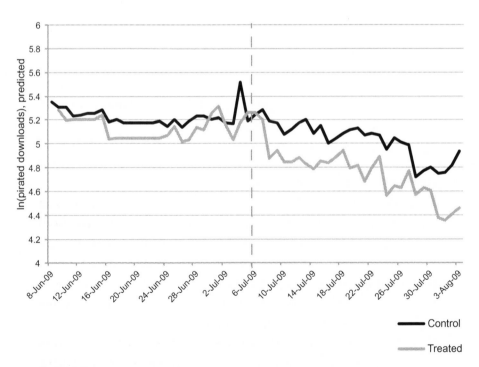

Figure 6.5
Piracy levels before and after ABC added its television content to Hulu.

for the treated ABC programs against piracy levels for programs in the control group during the four weeks before and after ABC added its content to Hulu.[24] As the figure shows, piracy levels in the control and treatment groups were similar in the four weeks preceding July 6, but immediately after ABC's content was added to Hulu there was a significant decrease in piracy of treatment-group titles relative to control-group titles, which in percentage terms corresponds to about a 16 percent reduction in piracy caused by making the content available on Hulu.

In short, just as Amazon can use the expectation of superior service, convenience, and reliability to differentiate its products from those of lower-price competitors, firms in the entertainment industry can use reliability, convenience, and the personal satisfaction of not stealing content to differentiate their legal offerings from "free" pirated content.

● ● ●

But firms in the entertainment industries have another important strategy at their disposal for competing with piracy. Not only can they control the differentiation of their own offerings (as Amazon can do in its competition with other booksellers); they also can reduce the convenience, quality, and reliability of their pirate-competitors' products by partnering with governments to make it more difficult or more legally risky to consume pirated content.

The ability to use legal measures in the fight against piracy raises an important final question: Can these sorts of anti-piracy interventions cause people to switch from piracy to legitimate purchasing? We have studied this question as it applies to three very different anti-piracy interventions: notice-sending programs targeting the demand side of piracy, and site shutdowns and site-blocking efforts targeting the supply of piracy. In each case we have found that the answer is Yes.

In 2012, for example, we studied the effect of a notice-sending anti-piracy law passed in France in 2009, commonly referred to as HADOPI. The fact that the law affected only French consumers allowed us to examine the law's effect on iTunes sales patterns in France versus those in a control group of statistically similar countries. This control-treatment approach allowed us to isolate the effect of the law on French consumers from other unrelated changes (e.g., seasonality or the effect of Apple product releases) that would have affected both French consumers and consumers in the control group. The data showed that the French anti-piracy law caused a 20–25 percent increase in music sales in France relative to the control group.[25] We also found that the increase in sales was larger for heavily pirated genres (rap, hip hop, and rock) than for less heavily pirated genres (classical, jazz, folk, and Christian), which is consistent with what one would expect given that rap, hip hop, and rock consumers were more heavily affected by the law than consumers of less pirated genres were. Apparently, making it legally more risky to commit piracy caused some consumers to switch from pirated content to legal consumption.

In 2014 we studied a different approach to fighting piracy: shutting down entire piracy sites. We focused on the shutdown of a piracy cyberlocker called Megaupload.com. Before it was shut down, Megaupload housed more than 25 petabytes of largely copyright-infringing content and accounted for 4 percent of all traffic on the Internet.[26] It was shut down in January of 2012 in a complicated legal and law-enforcement operation led by the US Department of Justice, which for our purposes represented a natural shock to the ease of accessing pirated content online.

However, because Megaupload was shut down worldwide, we couldn't compare sales in affected countries against sales in unaffected countries, as we had in our study of the HADOPI law. We knew, however, that Megaupload was far more popular in some countries than in others, and that fact allowed us to develop a hypothesis: If shutting down Megaupload had an effect on user behavior, its effect should be far larger in countries in which Megaupload was popular than in other countries. And that is exactly what we found.

For the study, we analyzed digital sales of movies in twelve countries before and after the shutdown. We found that the increase in sales relative to historical norms was far larger in countries that were heavier users of Megaupload before its shutdown than in other countries.[27] Using that result, we were able to conclude that shutting down Megaupload caused an increase in digital movie sales of 6.5–8.5 percent for the studios in our dataset. Apparently, shutting down a major piracy site can also cause some consumers to switch from pirated content to legal consumption.

But shutting down piracy sites is complicated, particularly when a site is housed outside a country's borders. For example, the Megaupload shutdown required precise coordination among law-enforcement agencies in nine countries and involved the simultaneous execution of twenty separate search warrants.[28] For this reason, many countries have opted for a simpler approach: blocking Internet users' access to sites. Site-blocking laws work by requiring Internet service providers in a particular country to block users' access to piracy-enabling sites that have been identified in court proceedings. However, savvy users can find ways to get around the blocks. That raises a question: If site blocking is imperfect, does it have any effect on consumer behavior? We asked that question in 2015 in a study of the effect of site-blocking regulations in the United Kingdom. For the purposes of the study, we obtained data documenting consumers' visits to both blocked sites and legal sites. We used a strategy similar to the one we had used to study Megaupload: If site blocking affects the behavior of consumers, it should affect heavy users of the blocked sites much more than light users.[29] That, again, is exactly what we found, but this time with a twist. We found that the shutdown of the popular piracy site The Pirate Bay in May of 2012 had no effect on visits to legal sites; former users of the blocked site simply shifted their visits to other piracy sites. However, when nineteen sites were blocked simultaneously in late 2013, we found a 12 percent increase in visits to legal movie-streaming sites. As was expected, the increase was far higher (23.6 percent) among heavy users of the blocked sites than among lighter users (3.5 percent). This suggests that blocking doesn't have to be perfect to be effective. Some former pirates will switch to legal channels if it becomes sufficiently hard for them to find pirated content.

In short, Steve Jobs was right. You can't stop piracy; you have to compete with it. What our data have taught us, however, is that this competition can take two forms. You can compete against free piracy by making the paid version easier to use, more convenient, and more reliable; or you can compete against piracy by making pirated content harder to use, less convenient, or less reliable. This is good news for content creators, and if piracy were the only threat facing the creative industries, these anti-piracy and pro-market strategies might go a long way toward ensuring the health and survival of the majors' current business models in the years ahead. But as we've already shown in our discussion of the long tail, piracy *isn't* the only major threat affecting the entertainment majors. In the next chapter, we'll talk about another one: new opportunities that artists have to create and distribute content, and the resulting explosion in "self-produced" content.

Appendix

Table 6.1 lists peer-reviewed journal articles finding no statistical effect of piracy. Table 6.2 lists peer-reviewed journal articles finding that piracy harms sales.

Table 6.1
Peer-reviewed journal articles finding no statistical impact of piracy.

	Primary data	Result
Oberholzer-Gee and Strumpf (2007, *J. of Political Economy*)	2002 OpenNap music downloads, 2002 US sales of popular albums	"[F]ile sharing has had no statistically significant effect on purchases of the average album in our sample."
Smith and Telang (2009, *MIS Quarterly*)	2005–06 Amazon DVD sales ranks and BitTorrent movie file downloads	"[T]he availability of pirated content at [television broadcast] has no effect on post-broadcast DVD sales gains.
Andersen and Frenz (2010, *J. of Evolutionary Economics*)	2006 survey of Canadian customers' file-sharing and CD purchasing behavior	There is "no (statistical) association between the number of P2P files downloaded and CD album sales."

Table 6.2

Peer-reviewed journal articles finding that piracy harms sales.

	Primary data	Result
Hui and Png (2003, *Contrib. to Economic Analysis & Policy*)	1994–1998 IFPI worldwide CD sales data and physical piracy rates	"[D]emand for music CDs decreased with piracy, suggesting that 'theft' outweighed the 'positive' effects of piracy."
Peitz and Waelbroeck (2004, *Rev. of Econ. Res. on Copyright*)	1998–2002 worldwide CD sales, IPSOS survey data for piracy downloads	"The implied loss of CD sales due to MP3 downloads is –20% for the period 1998–2002."
Zentner (2005, *Topics in Economic Analysis and Policy*)	1997–2002 country-level data on music sales and broadband usage	"Countries with higher internet and broadband penetration have suffered higher drops in music sales."
Stevens and Sessions (2005, *J. of Consumer Policy*)	1990–2004 consumer spending on cassette tapes, LPs, and CDs	"[T]he proliferation of peer-to-peer file sharing networks since 2000 has led to a *significant* decline in music format sales."
Bounie et al. (2006, *Rev. of Econ. Res. on Copyright*)	2005 survey of movie piracy and purchases from French universities	"[Piracy] has a strong [negative] impact on video [VHS and DVD] purchases and rentals" but statistically no impact on box office revenue.
Michel (2006, *Topics in Economic Analysis and Policy*)	1995–2003 US BLS micro Consumer Expenditure Survey data	"The relationship between computer ownership and music purchases weakened" due to piracy, potentially reducing CD sales by 13 percent.
Rob and Waldfogel (2006, *J. of Law and Economics*)	2003 survey of US college students' piracy and purchase behavior	"[E]ach album download reduces purchases by 0.2 in our sample, although possibly by much more."
Zentner (2006, *J. of Law and Economics)*	2001 survey of European music purchase and piracy behavior	"[Piracy] may explain a 30 percent reduction in the probability of buying music."
Bhattacharjee et al. (2007, *Management Science*)	1995–2002 Billboard 100 chart rankings, WinMX file-sharing post 2000	P2P file-sharing technologies have resulted in "significantly reduced chart survival except for those albums that debut high on the charts."

Table 6.2 (continued)

	Primary data	Result
DeVany and Walls (2007, *Rev. of Industrial Organization*)	Box-office revenue and supply of pirated content for an unnamed movie	"[Piracy] of a major studio movie accelerated its box-office decline and caused the picture to lose about $40 million in revenue."
Hennig-Thurau, Henning, Sattler (2007, *Marketing Science*)	2006 survey of German movie purchase and piracy intentions	Piracy causes "substantial cannibalization of theater visits, DVD rentals [and] purchases responsible for annual revenue losses of $300 million in Germany."
Rob and Waldfogel (2007, *J. of Industrial Economics*)	2005 survey of U. Penn. students' movie purchase and piracy behavior	"[U]npaid first [piracy] consumption reduces paid consumption by about 1 unit."
Liebowitz (2008, *Management Science*)	1998–2003 census data on broadband Internet use and music purchases	"[F]ile sharing appears to have caused the entire decline in record sales [observed from 1998 to 2003]."
Bender and Wang (2009, *International Social Science Rev.*)	1999–2007 country-level annual recorded music sales	"For a one percent increase in piracy rate, music sales declined about 0.6 percent."
Danaher et al. (2010, *Marketing Science*)	2007–08 BitTorrent downloads of television torrents	"[T]he removal of NBC content from iTunes resulted in an 11.4% increase in piracy for its content"
Waldfogel (2010, *Information Economics and Policy*)	2009–2010 survey of Wharton students' music piracy and purchases	"[A]n additional song stolen reduces paid consumption by between a third and a sixth of a song."
Bai and Waldfogel (2012, *Information Economics and Policy*)	2008–09 survey of Chinese university students' movie behavior	"[T]hree quarters of [Chinese students'] movie consumption is unpaid and … each instance of [piracy] displaces 0.14 paid consumption instances."

Table 6.2 (continued)

	Primary data	Result
Danaher et al. (2013, *J. of Industrial Economics*)	2008–2011 iTunes music sales in France and other European countries	The HADOPI anti-piracy law "caused iTunes music sales to increase by 22–25% [in France] relative to changes in the control group [countries]."
Hong (2013, *J. of Applied Econometrics*)	1996–2002 survey data from US BLS Consumer Expenditure Survey data	"[F]ile sharing is likely to explain about 20% of the total sales decline during the Napster period, mostly driven by downloading activities of households with children aged 6–17."
Danaher and Smith (2014, *International J. of Industrial Organization*)	2011–2013 sales and rental data for movies in 12 European countries from 3 major motion-picture studios	"The shutdown of Megaupload and its associated sites caused digital revenues for three major motion picture studios to increase by 6.5–8.5%."
Ma et al. (2014, *Information Systems Research*)	Box office revenue for all movies whose wide release occurred between February 2006 and December 2008.	"Pre-release [movie] piracy causes a 19.1% decline in revenue compared to piracy that occurs post release."
Adermon and Liang (2014, *J. of Economic Behavior & Organization*)	Digital and physical music sales in Sweden, Norway, and Finland, 2004–2009.	The IPRED copyright reform measure in Sweden "increased music sales by 36% in during the first six months [after it was implemented]. Pirated music therefore seems to be a strong substitute to legal music [purchases]."

7   Power to the People

Until now, those of us in the television and film business had been able to wait for the talent to find us. We had the keys to the kingdom, and folks needed to bring us their stories if they wanted to find a route to an audience. But now things are changing, and changing fast.
Kevin Spacey, 2013 James MacTaggart Memorial Lecture, *Guardian* Edinburgh International Television Festival

As we discussed in chapter 2, there was a time when writers, musicians, and actors were dependent on the "majors." Signing a contract with a major publisher, label, or studio was the only way to get access to the funding and the production expertise necessary to create content, and to the scarce promotion and distribution channels necessary to distribute this content to an audience. But things are now changing fast, for a variety of reasons.

First, the cost of creating content has gone down to the point where professional quality production is now available to the masses. For many types of content, artists no longer need expensive equipment to practice their trade. The cinematographer Kieran Crilly, for example, shot the Academy Award–winning documentary *The Lady in Number 6* with a Canon 5D Mark III, a camera that sells for a few thousand dollars on Amazon.[1] And many major movie releases—including the 2010 and 2011 Academy Award–winning movies for Best Editing—were edited with Final Cut Pro, a software package that sells for $300.[2]

Second, the cost of production facilities has declined to the point where professional-quality facilities are now within reach of many artists. Any YouTube artist with 5,000 or more subscribers can join the YouTube Partner program and gain access to YouTube Spaces—production facilities, located in Los Angeles, New York, Tokyo, London, Berlin, Mumbai, and São Paulo, that give YouTubers access to professional-quality production and editing equipment and to classes on makeup, design, and videography.

Third, it's now increasingly feasible to hire freelance talent to push a project forward, as the romance novelist Barbara Freethy did when she wanted to release several of her out-of-print novels internationally. Instead of using the staff at a publisher, she used Elance.com. The online marketplace for providers of freelance professional services allowed her to hire translators and editors for her German, Spanish, and French editions.[3]

Beyond production, some creators are using technology to crowd-source content development. That's what Neelesh Misra, known as "The Pied Piper of Indian Radio," does for the radio program he hosts, *The Idiot Box of Memories*. The program has 42 million radio listeners in India, and has recently broadened its reach to include Facebook and YouTube.[4] Broadcast each weekday, it centers on a 15–20 minute story about everyday life in India. Finding enough ideas to support 200 or so new stories a year is a huge task, naturally. Where does Misra get his stories? From listeners, of course. He sponsors writers' clubs across India where members write, and then read, thousands of stories, and choose the best ones for his show.

New technological developments have also democratized access to sales platforms such as Apple's iBookstore and Amazon's Kindle Direct Publishing program (for books), Bandcamp, Pledgemusic, and Amazon's Artist Central (for music), and YouTube's Partner program (for motion-picture content). And the increased flexibility offered by digital distribution is allowing creators to break the creative mold of traditional albums, movies, television shows, and books. For example, *The Saint*, Oliver Broudy's 28,000-word memoir about traveling with a collector of Gandhi-related memorabilia, was too long for magazine editors and too short to be a print book, but it became a brisk seller on Amazon's Kindle Singles platform.[5]

Artists are also making use of expanded funding opportunities. In 2012, Seth Godin raised $40,000 in less than four hours,[6] and raised $280,000 in all,[7] from Kickstarter supporters willing to fund production of his book *The Icarus Deception*. A year later, the creators of the television series *Veronica Mars* raised $2 million in less than ten hours,[8] and a total of $5.7 million,[9] in a Kickstarter campaign to fund production of a feature-length movie after the television show wasn't renewed for a fourth season by the UPN and CW networks.[10]

What does this mean for the entertainment industries? Let's begin with what it doesn't mean. It doesn't mean that studios, publishers, and labels no longer serve a purpose. The majors will continue to provide artists with support for production facilities and expertise, promotional funding, and access to distribution channels. But we believe that these changes threaten the majors' long-term power and profitability, because the growth of do-it-yourself production and distribution will change the majors' influence in their relationships with four important constituencies: artists, consumers, existing business partners, and distributors. We'll explore each of these relationships below, beginning with artists', who now have new opportunities to create content and reach their audience without the help of industry gatekeepers.

Artists

What if you and a friend wanted to combine your interests in songwriting and improvisational comedy to develop a series of imaginary musical duels between significant historical personalities? Or what if you had a passion for writing romantic vampire novels and wanted to reach young adult readers? Or what if you wanted to bring your vision of choreographed dance and hip-hop violin music to the masses? Until recently, unless you were a well-known artist, you would have had little chance of convincing industry gatekeepers to take a chance on your quirky creative visions. But today you might be able to realize your dream. Each of the above scenarios, in fact, corresponds to a real-life story.

Let's start with Peter Shukoff and Lloyd Ahlquist, who have made the idea of dueling musicians into a reality and have become an Internet sensation. They teamed up in the late 1990s after Ahlquist recruited Shukoff to join his improv-comedy group, Mission IMPROV-able. The group toured college campuses and small comedy clubs, and one of its routines involved improvised rap contests between historical figures suggested by the audience. By 2009, Shukoff had his own YouTube channel of comedy routines and had come to believe that the group's rap battle sketches were ideal for a pre-recorded format, which would allow the duo to spend more time researching their characters, staging their interactions, and creating sophisticated audio and visual effects. He and Ahlquist decided to bring their vision to the market.

How did they do it? First, let's consider their funding. In the late 1990s professional video production required specialized equipment and expensive facilities, but by 2009 smartphone users carried high-definition video cameras around in their pockets, and sophisticated video editing and audio mixing software could be purchased for a few hundred dollars. As a result, Shukoff and Ahlquist were able to produce the first three videos in their series on a budget of $50.[11]

Similarly, by 2009 new distribution options had been introduced, and Shukoff and Ahlquist didn't have to restructure their creative vision to fit within a standard 30-minute television broadcast slot, or convince television executives to take a chance on their niche concept. YouTube, founded in 2005, allowed its users to upload videos of virtually any length.

What about production expertise and creative support? Shukoff honed his video-production skills by watching a series of videos on the topic—videos he had discovered on YouTube.[12] And for creative support, instead of hiring a team of writers, Shukoff asked subscribers to his YouTube channel for ideas. When a fan proposed a competition between John Lennon and Bill O'Reilly, Epic Rap Battles of History was born.

The first battle, which featured Shukoff (a.k.a. NicePeter) as Lennon and Ahlquist (a.k.a. EpicLloyd) as O'Reilly, debuted on Shukoff's

YouTube channel on September 26, 2010, and garnered 150,000 views in the first two weeks. The sketch closed with the tagline "Who won? Who's next? You decide." Ideas came flooding in, and "epic" word of mouth allowed the second video, which featured Darth Vader and Adolf Hitler, to generate a million views in the five days after its November 2010 debut.[13]

The popularity of Shukoff and Ahlquist's channel increased, and by 2015 Epic Rap Battles of History was in its fourth season, had featured fifty battles, and had generated more than 1.7 billion views. The ERB YouTube channel has 12.2 million subscribers, making it the sixteenth-most-popular channel on YouTube.[14] Beyond YouTube, all of ERB's battles are available for purchase on iTunes. In fact, ten of them—including those pitting Darth Vader against Hitler, the Mario Brothers against the Wright Brothers, Steve Jobs against Bill Gates, and Barack Obama against Mitt Romney—have been certified gold by the Recording Industry Association of America.[15, 16]

Let's now consider the case of Amanda Hocking, who has attained fame and riches by self-publishing a series of romantic vampire novels. As a teenager growing up in Austin, Minnesota, Hocking was a prolific author of young-adult paranormal fiction. But by age 25 she had little to show for her efforts other than seventeen unpublished novels and a stack of rejection letters from traditional publishing houses. In addition to her love for vampire stories, Hocking loved Jim Henson and the Muppets, and this is where things get interesting. In April of 2010, she learned that a Muppets convention was scheduled for November. Unfortunately, her $18,000-a-year job at a group home for the disabled was barely paying her bills and left her no disposable income for a trip from Minneapolis to Chicago to see the convention. Desperate to attend, and looking to raise $300 for travel and hotel costs, she decided to sell one of her books on Amazon. Surely in six months, she thought, she could make $300 through Amazon's self-publishing platform. And she was right. In her first six months on Amazon, she made $20,000—and in the next fourteen months she made another $2.5 million.[17]

Lindsey Stirling has a similar story to tell. She has made a name for herself online as a dancing hip-hop violinist.

A hip-hop violinist? A *dancing* hip-hop violinist? Really? "I auditioned for talent agencies, I went to agents, and no one could capture the vision I saw in my head," she told the *Washington Post*. "I kept being told it was not marketable at all," she said in another interview, "something we don't want to touch."[18] In 2007, working to pay her way through college and discouraged after discovering that (as she put it) even trying to make it in the music business would cost her "several hundred thousands of dollars,"[19] Stirling decided to put a video of her music onto YouTube—and she now has nearly 7 million subscribers and more than 1 billion views on her channel.[20] She has put out two albums, which spent a combined 127 weeks on the Billboard 200 chart, peaking at 23 and 2 respectively; and in 2015 she completed a fifty-five-city world tour, selling out such major venues as Red Rocks and the Central Park Summerstage.

The change here is obvious—and significant. Increasingly, artists are able to bypass the majors to reach their audiences, and some artists who have established themselves on their own will be able to negotiate better deals when contracting with the majors than they otherwise would have been able to do. Given these new options, many independent artists will choose to remain independent and will replace the packaged set of services provided by studios, labels, and publishers with à la carte services that meet their specific needs.

For example, Peter Shukoff and Lloyd Ahlquist now work with Maker Studios, a content producer specializing in short-form digital content, and say they chose Maker rather than a major studio for artistic reasons. "Their philosophy," Ahlquist has said, "is that in this day and age, and in the YouTube market, the individual vision needs to be pure and unique … . They provide structure, resources, and support, but almost no editing of content."[21]

Lindsey Stirling, instead of signing with a major music label, signed with Lady Gaga's agent, Troy Carter, who has encouraged her to remain an independent presence on YouTube. "She's getting more views on

YouTube than you'd be able to get from radio or performing on TV," Carter has said. "We want to guide her career in the way we would any other artist's career, by keeping it independent."[22] Carter also has said that she had plans to bring in a private distributor "to help out with the physical goods," which strikes us as a telling remark. It suggests that although the ability to control distribution was once considered a big part of a label's power, Carter and others now think of it as a commodity that one can buy independently.

Of course, not every artist will choose the independent route. Many independent artists will choose to "graduate" from independent status to a contract with a major label, studio, or publisher. The problem for the majors, however, is that when the artists do graduate they will have more options, and therefore more negotiating power—the sort of power that, in April of 2011, allowed Amanda Hocking to start a bidding war among major publishers that reportedly resulted in bids exceeding $2 million for the English-language rights to Hocking's four upcoming books.[23] That's pretty heady stuff for someone who just a year earlier was an unpublished, struggling author with a shoebox full of rejection letters, some of them from the same publishers who were now bidding for her services.

So far we have only talked about how these new options affect emerging artists, but these technological developments are also giving established artists new negotiating power with the majors and new opportunities to bypass the majors entirely. For example, after its contract with EMI expired in 2003, the band Radiohead decided not to renew it. "I like the people at our record company," the group's lead singer, Thom Yorke, said, "but the time is at hand when you have to ask why anyone needs one."[24] In 2007, Radiohead released their next album, *In Rainbows*, independently, using their website, Radiohead.com, to distribute content directly to fans. Instead of the normal fixed price of $10–$15 for the album, the band let their fans decide what to pay. Fans could download it entirely for free or could pay whatever they thought it was worth. What did Radiohead discover about their fans' generosity (and about the advantages of cutting out the middleman)? "In terms of digital income, we've made more money out of this record than out of all the other Radiohead albums put together," Yorke told David Byrne in a discussion with *Wired* magazine.[25]

In 2011 the comedian Louis C.K. made a similar discovery about the power of direct distribution by conducting an experiment. He described the experiment in a blog post: "If I put out a brand new standup special at a drastically low price ($5) and made it as easy as possible to buy, download and enjoy, free of any restrictions, how much money can be made by an individual in this manner?"[26] The answer? $1 million in the first 12 days, which, after paying for the costs of the video production and website development ($250,000), left the comedian with $750,000 in profit—$250,000 of which he distributed to his staff as a "big fat bonus," $280,000 of which he donated to various charities, and $220,000 of which he kept for himself.[27] Since then, he has released three other comedy shows on his site. *Live at the Comedy Store* (January 2015) garnered nowhere near as much press attention as the 2011 release, but, still, after just four days, the comedian said it was "selling better so far than anything else had this far in."[28]

However, the queen of self-distribution is J. K. Rowling, who negotiated a deal with her publisher that allowed her to retain digital rights to her Harry Potter books. The deal allowed her to create Pottermore.com, a digital storefront that is the exclusive seller of Harry Potter e-books. Even Amazon, a company that controls 90 percent of online e-book sales, had to bow to Rowling's power. When customers want to purchase Harry Potter e-books, Amazon refers them to Pottermore.com and takes a fee on the sale. This arrangement allows Rowling to maintain a direct connection with her customers and to build fan engagement and customer loyalty. Pottermore.com debuted with more than 18,000 words of previously unpublished content, allowing Harry Potter and his world, as Rowling has put it, "to live on in a medium that didn't exist when I started writing the books."[29] (The irony here is too delicious to pass over: Amazon, the company that has brought major publishers to their knees and is bent on transforming the book business, had to submit to a single author's demands to maintain direct control over her customers. "Schadenfreude," Philip Jones wrote in *The Guardian*, commenting on the situation, "doesn't even come close."[30])

Not all established artists are going to take the direct-to-consumer approach. However, even when they don't, these new options will give them more leverage in their negotiations with the majors, which will result in less profitable deals for the majors. This is problematic for those companies, because, as we pointed out in chapter 2, the entertainment industries rely on the fat profits generated by a small number of successful artists to subsidize the risk they carry on their lesser-known artists.

Consumers

In addition to changing the majors' relationships with artists (and the profitability of those relationships), technological developments have changed the majors' relationships with their customers, who now have many new forms of content to consume.

We have heard many in the industry dismiss the new self-produced content as "inferior"—amateur material, produced by "vanity press" authors and "cover band" musicians, that can't compete with the quality of what professionals produce. We believe this reveals a fundamental misunderstanding of the nature of competition. In a market economy, it is buyers, not sellers, who determine whether products are in competition. It doesn't matter how good the product is, or whether it conforms to established industry norms. If customers are choosing a product other than yours, you are losing to the competition. And a lot of evidence suggests that this is exactly what is happening. According to Nielsen Soundscan data, the share of the music business belonging to independent artists increased from 25.8 percent to 34.5 percent between 2007 and 2014, and independent artists now capture a larger share of the industry than any individual major label.[31] Publishing has seen a similar shift; according to Joel Waldfogel and Imke Reimers, the number of self-published titles increased by 300 percent from 2006 to 2012, surpassing the number of traditionally published titles.[32] Similarly, 2 million of the 3.5 million e-books Amazon sells come from independent authors on its Kindle Direct Publishing platform.

But the biggest shift in consumption has occurred in the motion-picture industry. Millennials, in particular, seem to be consuming self-produced content instead of the "professional" content produced by the majors, which, as the *Washington Post* recently reported, seems to be "increasingly for old people." Consider the following sobering statistics, drawn from a variety of recent sources and surveys: The median age of all television viewers in the 2013–2014 season was 44.4 years, and for the major networks it was 53.9 years—increases of 6 and 7 percent, respectively, from just four years earlier.[33] Young people's attendance at movie theaters fell by 40 percent from 2002 to 2012.[34] The ratings of prime-time live television shows for 18–49-year-olds fell by 50 percent from 2002 to 2011.[35] One in four millennials is a "cord cutter" (i.e., has chosen to stop subscribing to a cable service), and one in eight is a "cord never" (i.e., has never subscribed to a cable service).[36] TV viewing among 18–24-year-olds fell by 32 percent from 2010 to 2015 (but by only 1 percent among 50–64-year-olds).[37] In 2014, only 21 percent of 18–24-year-olds said that they "can't live without their TV," versus 57 percent of the broader population.[38]

Where are younger viewers going? Online. In 2014, for the first time, YouTube reached more 18–34-year-olds than any cable network,[39] and the number of 18–24-year-olds who said they "can't live without their smartphone" jumped to 50 percent, up from 22 percent in 2011.[40]

Business Partners

For the majors, the developments cited above present not only dangers but also opportunities. As we will discuss in more detail in chapters 10 and 11, we see many ways in which these firms might use online channels to distribute their content and create new connections with customers, and many ways they might leverage lower costs and new technologies in their businesses. But in many cases making the most of these new opportunities will put pressure on existing business models and partners—and coping with that pressure will be difficult.

The most obvious example of this sort of pressure comes from the new downstream distribution technologies themselves. The same technologies that are enabling artists to bypass industry gatekeepers to reach their audiences can also enable the established networks to deliver content when and where their consumers want it. This usually sounds like a great idea to the networks, but only until the popularity of these new distribution channels threatens the profitability of their existing channels. At that point, managers face a very difficult decision between their existing business model (which is often tied to their salaries and bonuses) and the new one.

New business opportunities may also complicate the majors' relationships with upstream business partners. Integrating user-generated content into a company's established business practices, for example, isn't as easy or as seamless as it might appear. In many cases, online fans have come to expect access and participation that are difficult for established companies to provide within their current business structure. ABC discovered that in 2008 when it purchased the rights to the successful online show *In the Motherhood*. The show began its life in 2007 as a parenting community and a Web series created "by moms, for moms, about moms." The premise of *In the Motherhood* was simple. As the show's site put it, mothers wanting to "sound off about universal experiences all moms share" (your child's worst meltdown, etc.) would log on and submit a brief real-life story, which might then be nominated by "the online mom community" to be "polished and incorporated into the series of webisodes."[41]

In its first online season, *In the Motherhood* had strong brand sponsorship from Suave and Sprint and attracted established talent, including the director Peter Lauer (*Arrested Development*, *Malcolm in the Middle*, *Chappelle's Show*) and the actors Jenny McCarthy, Chelsea Handler, and Leah Remini, who portrayed the "ever-hectic yet always humorous lives of three mom girlfriends." The episodes were available exclusively on the MSN site inthemotherhood.com, and in the first season they garnered 5.5 million views.[42] The show's online community on MSN became the fifth-most-visited online parenting site,[43] generating 3,000 submissions and 60,000 votes.[44]

Seeing a potential hit, ABC purchased the rights to *In the Motherhood* in September of 2008,[45] initially ordering thirteen episodes to air. On March 11, 2009, hoping to maintain a strong creative connection with the online community, ABC invited mothers to submit their best stories to the network's website.[46] That's when the Writers Guild of America got involved. The Writers Guild complained that ABC was, in effect, asking fans to generate stories for free, whereas ABC's contract with the union required that all writers working for the network are compensated as if they were WGA members. "This kind of call for submission is not allowed for in the contract," said the WGA spokesman Neal Sacharow. "It's not our goal to curtail experimentation, but people who do the work should get paid for it"—by which he meant a fee at least equal to the WGA's minimum of $7,000 per submission.

Two weeks later ABC pulled the request for audience submissions from its website and turned to in-house writers. But those writers' efforts couldn't match the energy of real stories and real audience participation, and the series' premiere attracted only a meager audience. Halfway into the first season, the network canceled the show.

Distributors

We believe that the volume of new content alone is bringing about changes in the relationship between content providers and their online distributors. Since the turn of the century, thanks to new technologies that have enabled artists to produce content themselves, we have witnessed an explosion in the amount of content produced. The number of new book titles, for example, went from 122,000 in 2000[47] to 3.1 million in 2010.[48] The number of new albums quadrupled during a similar period,[49] and 300 hours of video now are uploaded to YouTube *per minute*.[50] But how can consumers sift through these offerings and discover what they like?

In the past, the majors did the sifting and the discovery themselves, before consumers ever saw anything, and then decided what consumers would like. But that top-down model is now changing. Increasingly, discovery is occurring downstream, via distribution platforms that bring together masses of content, learn consumers' preferences, and recommend specialized material directly to consumers. Labels, studios, and publishers aren't well positioned to exploit that opportunity with their existing business models. Instead they have ceded this opportunity to a new breed of online distributors, among them Amazon, Netflix, YouTube, and iTunes. Why is that a threat? We'll take up that topic in our next chapter.

Those nerds are a threat to our way of life.
Stan Gable, *Revenge of the Nerds*

On August 31, 2007, Apple announced that NBC Universal had refused to renew its contract to sell television shows on the iTunes platform. The specific terms of the dispute are unclear, but most accounts indicate that Apple had declined to comply with three requests made by NBC: increased pricing flexibility on iTunes,[1] anti-piracy measures to make it harder for customers to load pirated content onto iPods,[2] and a share of Apple's revenue from iPod sales.[3] But beneath these specific demands lay a deeper problem. The iTunes movie and television stores were rapidly gaining market power, and the industry was worried. "Apple has destroyed the music business, in terms of pricing," Jeff Zucker, NBC Universal's CEO, told students that fall at Syracuse University's Newhouse School of Communication, "and if we don't take control, they'll do the same thing on the video side."[4]

NBC refused to renew its contract with Apple for three reasons. First, it felt it had the upper hand in the dispute. It was the top supplier of video content to Apple's iTunes store, reportedly accounting for 40 percent of Apple's video sales.[5]

Second, the timing was right. Apple was about to announce its first-generation iPod touch, with video playback capability, at a media event on September 5, 2007,[6] and certainly Apple was going to have more difficulty selling the device if consumers knew they wouldn't be able to find NBC's television shows on the iTunes platform. Commenting on the situation at the time, James McQuivey of Forrester Research said "Apple is totally dependent on NBC and the rest of the content creators to make their video-playback devices ... valuable."[7]

Third, NBC's customers were no longer dependent on iTunes to watch NBC programs online. In the absence of iTunes, they could still buy boxed sets of NBC series from a variety of retailers, they could stream selected shows on NBC.com, and beginning September 4 they would be able to purchase digital downloads of NBC programs from Amazon's Unbox digital-download service. And they would soon have another option. In November NBC was going to launch its own platform, called NBC Direct, which was designed to replicate some of the services provided by iTunes. NBC Direct would allow consumers to download selected television shows for free (with embedded commercials) and watch them on compatible Windows computers, and soon (in early 2008) on Macintoshes. A future version of NBC Direct, planned for mid 2008, would even allow consumers to pay for commercial-free downloads, just like they could on iTunes.[8]

From NBC's perspective, the bet seemed safe. If customers no longer could get NBC content on iTunes, they would simply find it on one of the many other legal digital platforms—and Apple, if it wanted to ensure the health of the iTunes movie store and the iPod product line, would soon have to relent. In addition, a precedent existed for removing content to discipline retailers. Just a year earlier, the Walt Disney Company used a similar tactic to win a dispute against the Target chain, one of its largest DVD retailers.[9] Unhappy that Disney had just decided to sell movies through iTunes,[10] Target had announced it would no longer stock many Disney titles, and was returning cases and cases of DVDs to Disney and removing Disney's promotional displays from their stores. But the move backfired. Late in November, Disney threatened not to supply Target with DVD copies of a big holiday release of a movie that Target really wanted to carry (*Pirates of the Caribbean: Dead Man's Chest*),[11] and Target backed down.

If Disney was able to convince Target to back down by threatening to withhold one movie, NBC was understandably confident that it could bring Apple to its senses by withholding 40 percent of iTunes sales. Apple, for its part, was equally confident that it would come out on top. When asked by *USA Today* what effect NBC's departure would have on Apple and iTunes, Steve Jobs responded dismissively: "Overall, it's zero."

Who was right? Did Apple need NBC more than NBC needed Apple, or had the tables turned? When NBC's customers weren't able to buy content on iTunes, where did they go? We explored the last question immediately after NBC pulled its content from iTunes, and what we found may surprise industry executives who think they still exert considerable power over online retailers, even in the rapidly evolving world of digital commerce. NBC, it turned out, didn't have the power it thought it had. Instead of shifting to Hulu, Amazon or other legal channels, its former iTunes customers turned to piracy in droves—and generally didn't come back. We came to that conclusion after collecting and studying data on changes in BitTorrent piracy and DVD sales for NBC programs and a control group of NBC's closest competitors: ABC, CBS, and Fox. We first looked at how piracy changed after NBC left iTunes on December 1, 2007. Our results, summarized in figure 8.1, showed that before December 1 piracy patterns for NBC content closely matched piracy patterns for ABC, CBS, and Fox content, but immediately after NBC removed its content from iTunes the piracy of NBC content shot up. Relative to piracy in the control group, piracy of NBC content increased by 11.4 percent.[12]

Although the percentage change in piracy is striking, the unit change is, in some ways, more striking. We found that the weekly increase in BitTorrent downloads for NBC content after December 1 was twice as large as the *total* number of weekly NBC sales on iTunes before December 1. Why? As we discuss in a paper published in the academic journal *Marketing Science,*[13] the most obvious explanation is that once former iTunes customers learned how to use BitTorrent, instead of just consuming the handful of episodes they would have paid for on iTunes, they were able to download entire seasons of NBC content for free.

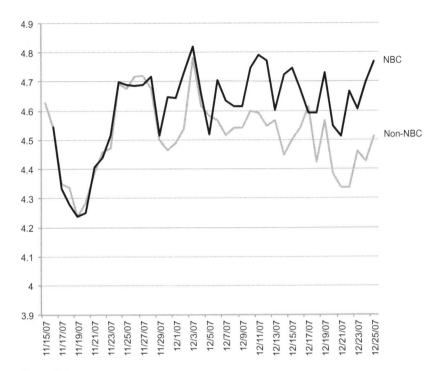

Figure 8.1
NBC vs. non-NBC piracy around December 1, 2007.

The fact that many former customers were now stealing content was bad news for NBC's strategy of forcing Apple back to the bargaining table. And the bad news kept coming. Not only did former iTunes customers switch to piracy in large numbers; the switch made it easier for other pirates to obtain NBC content on BitTorrent.

Why might increased demand for pirated content lead to increased supply? To answer that question, one must understand that the BitTorrent protocol is most efficient when a large group of users—known colloquially as a "swarm"—come together to upload and download the same piece of content. This characteristic explains why many seasons of older NBC shows were unavailable on BitTorrent before December 1. When people could purchase their content on iTunes, there was no need to download it on BitTorrent, and therefore there wasn't enough demand from pirates to support BitTorrent swarms for some of NBC's less popular shows. However, after December 1 we observed new BitTorrent swarms become available for 147 episodes of older NBC shows, including entire seasons of *Saved by the Bell* and *Xena: Warrior Princess*.[14]

That explains how NBC's piracy changed after it left iTunes. But what about legal consumption? More bad news. When we looked at our sales data, we observed no increase in sales of DVD boxed sets of NBC programs. We also discovered that the increase in streaming and downloads on NBC.com, Hulu.com, and Amazon's "Unbox" digital download service represented a tiny fraction of former iTunes consumption. In short, when consumers couldn't purchase on iTunes, most preferred piracy to consumption on other legal channels.

This obviously creates problems for NBC in its negotiations with Apple. But NBC's decision to leave iTunes may have also created difficulties for its competitors. Our data show that piracy of ABC, CBS, and Fox content increased by 5.8 percent immediately after December 1. We don't have a good counterfactual estimate for how non-NBC piracy would have changed if NBC had chosen to keep its content on iTunes, but the most likely explanation for the increase in piracy for NBC's competitors' is that once former iTunes consumers had learned how to pirate their NBC content through BitTorrent, many began to use BitTorrent to pirate ABC, CBS, and Fox content too. This inference is consistent with something we were once told by an executive at a competing network who described having observed an unusual decrease in their network's iTunes sales shortly after NBC stopped selling content on iTunes.

In short, the data showed that NBC's decision to withhold content from iTunes hurt NBC far more than it hurt Apple. NBC's customers switched to piracy in large numbers, and punishing iTunes did little to increase former iTunes customers' use of other legal channels. These two facts explain why NBC chose to bring its content back to iTunes for the start of the fall 2008 season and agreed to accept terms that were essentially the same as those it had rejected less than a year earlier.[15]

That decision, of course, raised another question: What happened to
NBC piracy when customers could again use iTunes?

The data showed that when NBC returned to iTunes, on September
8, 2008, NBC's piracy went down by only 7.7 percent—a decrease that
was far smaller, in both absolute and relative terms, than the increase in
December of 2007. Apparently, NBC's strategy backfired. Not only did
withholding content cause former iTunes customers to adopt piracy in
droves; once those customers learned how to use BitTorrent, it was hard
to get them to come back to iTunes.

But why? Why had the strategies the studios had used to successfully
discipline powerful brick-and-mortar retailers such as Target prove inef-
fective when applied to Apple? Why had power shifted away from studios, labels, and publishers and toward retailers?

• • •

We have observed, in the course of our research, that the Internet
doesn't threaten the entertainment industries only through long tail
processes, piracy, and increased artist power. It also threatens these
industries by allowing the creation of retail platforms on which power
tends to be concentrated in the hands of a small number of dominant
players.

At one level, of course, these new retail players have helped the
entertainment industries. Amazon sells a lot of books, and in many
ways it launched the e-book business. Many observers argue, too, that
iTunes has saved the record business, and that Netflix has opened
a new golden age of television production. But the dominance of
Amazon, iTunes, and Netflix in their respective markets means
that the studios, labels, and publishers can no longer resort to one of
their traditional strategies: playing retailers off of one another in
negotiations. In fact, in many ways the situation is now completely
reversed.

Consider the experience of publishers—particularly small ones—in
their negotiations with Amazon. Initially, small publishers were some

of the biggest beneficiaries of Amazon's service, because Amazon gave their books exposure that they weren't getting in major chain stores. But this dependence came at a cost. Small publishers became reliant on Amazon's services, and by 2004 Amazon was ready to exploit its power. As Brad Stone recounts in *The Everything Store*, Amazon's negotiating stance toward small publishers was known internally as the "Gazelle Project," so named because of Jeff Bezos' recommendation that Amazon "approach these small publishers in the way a cheetah would pursue a sickly gazelle."[16] Although Amazon's lawyers later changed the name to the less provocative "Small Publishers Negotiation Program," the concept was the same: Squeeze publishers that have nowhere else to turn. One way Amazon squeezed such publishers was by demanding cooperative advertising payments. In a brick-and-mortar setting, it is common for publishers and other entertainment companies to make "co-op" payments to retailers in exchange for featured shelf space in the retailer's stores and other promotional considerations. Amazon took that one step further by demanding that publishers make co-op payments in the form of 2–5 percent of their gross sales for the privilege of, among other things, appearing in Amazon's search results.[17]

That was too much for one small publisher. Dennis Johnson, the co-owner of Melville House, decided to stand up to the bully. "'Fuck you' was my attitude," he said in a 2014 *New Yorker* article. "They're bluffing—I'm going to call their bluff."[18]

Unfortunately for Johnson, Amazon wasn't bluffing. On April 1, 2004—the day after Johnson took his story to *Publishers Weekly* and openly criticized Amazon's "bullying" tactics—Amazon removed the "buy" button from each of Melville House's titles. Facing already slim margins in a competitive business, and unable to recover lost Amazon sales through other retail channels, Melville House agreed to the co-op payment, and the "buy" buttons reappeared.[19] "How is this not extortion?" Johnson would later ask in an interview with the *New York Times*. "You know, the thing that is illegal when the Mafia does it."

Melville House isn't alone in receiving offers it can't refuse from a partner it can't live without. Amazon reportedly makes co-op payments a regular part of negotiations with publishers, both big and small, and its demands appear to have grown over time: from payments of 2–5 percent of gross sales in 2004[20] to reported demands of 5–7 percent of gross sales from large publishers in 2014, and as high as 14 percent from the smaller "gazelles."[21] Philip Jones, the editor of the trade magazine *The Book Seller*, memorably captured the double-edged nature of Amazon's role in the book industry. "The worst thing that could happen [to book publishers] would be for Amazon to go away," he told the BBC. "The second worst thing would be for it to become more dominant."[22]

It isn't only in publishing that dominant online retailers are both a blessing and a curse, of course. In music, iTunes is widely believed to have saved the business from the threat of Napster and from the labels' reliance on CD sales. "With the introduction of the iTunes software and other platforms," Cary Sherman, the chairman of the Recording Industry Association of America, observed, "Apple made it once again easy and accepted to pay for music."[23] But at the same time, Apple's power gave it the ability to dictate terms, even to the powerful major labels. In *Appetite for Self-Destruction: The Spectacular Crash of the Record Industry*, Steve Knopper describes what happened after EMI discovered that Apple was selling Coldplay's *A Rush of Blood to the Head* for $11.88 rather than for $12.99 (the price on which the companies had agreed). When EMI called Apple to complain, the Apple representative, instead of offering to change the price, reportedly said "Okay, you want us to take it down?"

The entertainment industries also have to consider the role that Amazon, Netflix, and similar services now play in generating valuable exposure for content, and, subsequently, market information about who is buying the content. For example, in 2009, Angela Bromstad, the president of NBC Universal's Television Studio, described iTunes as having played an important part in the success of the NBC series *The Office*. The data generated by iTunes, she said, had given NBC "another way to see the true potential [of the show] other than just Nielsen Media Research [data]." Without the iTunes data, she added, "I'm not sure we'd still have the show on the air."[24] Similarly, Josh Sapan, the president and CEO of AMC, claimed that allowing viewers to watch the first four seasons of *Breaking Bad* on Netflix helped produce a 200 percent increase in AMC's viewership in the show's fifth season.[25]

To understand the breadth of the problem, consider the remarkable statistics in table 8.1. According to the available data, in the United States each of today's leading online retailers of books, music, and movies has a larger market share than the *combined* market share of the two largest retailers in the comparable brick-and-mortar markets.

The top online distributors are more powerful than their brick-and-mortar counterparts not only in terms of local market concentration but also in terms of global reach. In a brick-and-mortar world, market leadership in one country rarely extends to the rest of the world. For example, Wal-Mart, a dominant brick-and-mortar retailer in the United States, has little market power elsewhere. In contrast, the digital distributors that dominate the US market also hold dominant positions in markets all over the world.

Back in the days when the Internet produced only a small fraction of total sales, this sort of market power represented little more than a nuisance relative to the far more profitable (and far less concentrated) brick-and-mortar distribution channels. But since digital sales exceeded physical sales for movies in 2008,[26] for books in 2012,[27] and for music in 2014,[28] powerful online retailers are causing major headaches for all of the entertainment industries.

Something very interesting has happened. For decades, as we explained in chapter 2, leading publishers, labels, and studios relied on barriers to entry and economies of scale to maintain control over their industries. But now the balance of power has shifted, and the big online retailers are playing—and winning—the very same game.

• • •

Table 8.1
Brick-and-mortar market share vs. Internet market share for books, music, and motion pictures.

	Market share	
	Brick-and-mortar	Internet
Books	Print books, 2013[a]: Barnes & Noble and Borders 22–23% combined	Print books, 2013[b]: Amazon 64% E-books, 2014–15[c]: Amazon 64–67%
Music	CD sales, 2000[d]: Best Buy 18% Wal-Mart 16%	Digital downloads, 2015[e]: iTunes 80–85% Music streaming (US) Spotify 86%
Motion pictures	DVD Sales, 2005–06: Wal-Mart 30–40%[f] Target 15%[g]	All videos streamed online, 2012[h]: YouTube 63% All movies streamed and downloaded online, 2010[i]: Netflix 61% DVD sales, 2005[j]: Amazon 90% Digital movie downloads, 2012[k]: iTunes 65% (movies), 67% television) Digital movie rentals, 2012[l]: iTunes 45%

a. See *The Book Publishing Industry*, third edition, ed. A. Greco, J. Milliot, and R. Wharton (Routledge, 2013), p. 221; also see http://www.publishersweekly.com/pw/by-topic/industry-news/bea/article/62520-bea-2014-can-anyone-compete-with-amazon.html.

b. See http://www.publishersweekly.com/pw/by-topic/industry-news/bea/article/62520-bea-2014-can-anyone-compete-with-amazon.html. The report also notes that the Internet accounts for 41% of all print book sales. A separate article (http://www.forbes.com/sites/jeffbercovici/2014/02/10/amazon-vs-book-publishers-by-the-numbers/) reports that e-books make up 30% of all book sales.

c. See http://www.wsj.com/articles/e-book-sales-weaken-amid-higher-prices-1441307826. Other reports place Amazon's market share of e-books between 65% (http://www.forbes.com/sites/jeffbercovici/2014/02/10/amazon-vs-book-publishers-by-the-numbers/) and 67% (http://www.publishersweekly.com/pw/by-topic/industry-news/bea/article/62520-bea-2014-can-anyone-compete-with-amazon.html, http://www.thewire.com/business/2014/05/amazon-has-basically-no-competition-among-online-booksellers/371917/)

Table 8.1 (continued)

d. Ed Christman, "Best Buy Acquires Musicland Chain," *Billboard*, December 2000, pp. 1 and 82.

e. http://www.wsj.com/articles/apple-to-announce-new-music-services-1433183201

f. In 2005, Edward Jay Epstein placed Wal-Mart's market share of DVDs at 30% (http://www.slate.com/articles/arts/the_hollywood_economist/2005/12/hollywoods_new_year.html), and NPD Group placed it at 37% (http://variety.com/2005/biz/features/store-wars-1117932851/). In 2006, the *New York Post* reported that Wal-Mart's market share of DVDs was 40% (T. Arango, "Retail-iation: Wal-Mart Warns Studios over DVD Downloads," September 22).

g. In 2006, *Wall Street Journal* reported Target's market share of brick-and-mortar DVD sales at 15% (S. McBride and M. Marr, "Target, a Big DVD Seller, Warns Studios over Download Pricing," October 9, http://www.wsj.com/articles/SB116035902475586468).

h. Nielsen reported that YouTube served 16.5 billion of the 26.2 billion video streams served in May of 2012 (http://www.nielsen.com/us/en/insights/news/2012/may-2012-top-u-s-online-video-sites.html).

i. The *Hollywood Reporter*, citing results obtained by the NPD Group, reported that Netflix had a 61% share of all movies downloaded or streamed on the Internet, Comcast was second at 8%, and iTunes had only 4%.

j. J. Netherby, Amazon.com Dominates in Online DVD sales, Reed Business Information, Gale Group, Farmington Hills, Michigan.

k. https://www.npd.com/wps/portal/npd/us/news/press-releases/the-npd-group-apple-itunes-dominates-internet-video-market/

l. https://www.npd.com/wps/portal/npd/us/news/press-releases/the-npd-group-apple-itunes-dominates-internet-video-market/, https://www.npd.com/wps/portal/npd/us/news/press-releases/the-npd-group-as-digital-video-gets-increasing-attention-dvd-and-blu-ray-earn-the-lions-share-of-revenue/

What are the new barriers to entry and economies of scale that online retailers are taking advantage of, and do they represent a long-term threat to the established power structures in the entertainment industries? To consider these questions, in the remainder of this chapter we'll focus on the barrier to entry and economies of scale posed by four important aspects of online competition: consumer search and switching costs, platform lock-in, information bundling, and online platform development.

Consumer Search Costs and Switching Costs

Remember how Robert Kuttner, in 1998, described the Internet as "a nearly perfect market?" Less than a year later, David Shaw, the head of D. E. Shaw, the hedge fund for which Jeff Bezos worked before founding Amazon, made a very different observation about that very same market. "While it is true that if all you want to do is to put up something for sale, the barriers of entry are extremely low on the Internet," he told the *New York Times*, "if you actually want to sell a lot of that stuff, they're quite high and getting higher all the time."[29]

Shaw's observation dovetails with the academic consensus on this subject, which we discussed in a 2000 book chapter co-authored with Erik Brynjolfsson and Joe Bailey. In the literature we reviewed, we found that for most consumer markets, scholars have rejected the idea of the Internet as a nearly perfect market for two main reasons.[30] The first has to do with the costs in time and in cognitive effort that consumers face when searching online. Although information is usually easy to find online, too much of it can easily overwhelm consumers, most of whom are, for lack of a better word, lazy. On the whole, they don't like to compare prices, or to think too hard about different competing offers, or to take the time to learn how to use unfamiliar websites.[31] In fact, they turn out to be willing to pay several dollars more for the convenience of buying from retailers that allow them to avoid those tasks.[32]

The second reason has to do with uncertainty. In a brick-and-mortar world, you don't have to worry much about a bookstore's quality. You hand the salesperson the money, and the salesperson hands you the book. But online, reliability matters. Will an unknown retailer deliver your book on time, or at all? Will it be reasonable about returns? Will it sell your information to advertisers, or send you spam e-mail? It's hard to know—and that's the point. Some aspects of retailer quality are more important, and harder to evaluate, online than they might be in a brick-and-mortar setting. And when a retailer's quality is important and is difficult to evaluate, consumers tend to deal with a retailer they already know, or to gravitate toward the popular retailers other consumers are using.

Personalized recommendations can also introduce switching costs for consumers. The more a site learns about its consumers' preferences, the more accurately it can recommend products on the basis of their specific tastes. This, in turn, creates an important barrier to entry for new firms that don't have the customer data necessary to replicate these personalized recommendations.

The overarching point here is that consumer search costs and switching costs create a formidable barrier for new entrants to established online markets. And the barriers to capturing consumers' attention and trust are rising all the time.

Platform Lock-in

The second barrier, which is related to the first, becomes particularly important when content is digital: Consumers value the simplicity of having all their digital content on a single platform rather than spread out over many platforms. Who wants to deal with the hassle of learning how to use a number of different platforms to watch movies, remembering which content is on which platform, or keeping track of the different rights and permissions offered by different services?[33]

Beyond the search and switching costs described above, the technology itself can create a lock-in between digital content and digital platforms, resulting in a new barrier to entry. Digital-rights-management encoding often limits purchased content to a particular distributor's ecosystem. Courts and legal scholars have argued that this raises significant antitrust concerns, in that once consumers own multiple iTunes movies, or multiple Kindle books, it is much more difficult for a new entrant (e.g., Apple's iBookstore, Barnes & Noble's Nook, Amazon Instant Video, or Google Play) to gain a foothold in the market.[34]

Information Bundling

Digitization makes it easier and more profitable for producers to sell entertainment goods in large bundles than what would be possible with physical goods, and bundling information in this way creates significant economies of scale. In the extreme, it can lead to a single "winner-take-all" outcome for the company with the largest bundle.[35]

The economic rationale behind bundling is similar to the rationale for the price-discrimination strategies we described in chapter 3. When products are sold individually and different consumers assign radically different values to the products (for example, *Breakfast at Tiffany's*, *Scream*, *Legally Blonde*, and *Hoop Dreams*) economic theory says that producers should use price discrimination strategies to maximize their profit. But, when goods are sold a la carte, for price discrimination to be most effective, sellers will need to accurately predict each consumer's value for each product, and then set the price for each consumer to that value.

Bundling multiple products allows a seller to do this very efficiently. The more products the seller has in a bundle (say, the 10,000 or so shows on Netflix), the better the seller can predict the average value of the bundle across consumers. Not everyone assigns the same values to the same movies, but in a large bundle the differences in the individual values average out. And if a seller can accurately predict the average value its consumers are willing to pay for a bundle of products, all that is left is to set a price just slightly below that value, extracting all the value from the consumers.

On the consumer side, the logic is even simpler. The more products a seller can offer customers in a single bundle, the more convenient the seller's offering will be for consumers, the more consumers will be willing to pay, and the less consumers will be distracted by competing offers.

But this means that when two bundlers compete for the same set of customers, the firm with the larger bundle will be able to do a better job of predicting an individual consumer's valuation of the bundle, and thus the firm with the larger bundle will be more profitable than the firm with the smaller one. And similar advantages extend to upstream competition. According to Yannis Bakos and Erik Brynjolfsson, "larger bundlers are able to outbid smaller ones" when licensing content because firms providing a larger bundle know that they can "extract more value from any given good" than a smaller bundler could.[36]

Platform Development

Scaling a brick-and-mortar business up to serve more customers typically involves leasing additional space, buying new stock, and hiring additional employees—an expensive proposition. Online, once the front-end and back-end systems are in place, scaling those systems up to serve more customers is relatively easy. The main barrier to entry, it turns out, is the initial investment in reliable and efficient systems.

NBC discovered both the cost and the complexity of front-end and back-end development in its launch of the NBC Direct platform, as we discussed at the beginning of this chapter. The platform never quite made it out of beta stage, and it was almost universally panned by reviewers. *Wired* described the experience of using it as "sheer frustration," and Arstechnica.com wondered if "NBC's new strategy with NBC Direct is to make it so unpleasant to watch the shows offline that users are driven to Hulu by default. Or BitTorrent."[37]

NBC isn't the only company that has experienced difficulties when rolling out technological platforms. The entertainment industries in general have plenty of experience with platform failure. Remember Pressplay and Musicnet, the recording industry's answers to iTunes? A more recent example was HBO's effort to develop back-end services for the HBO Go streaming platform. According to an article in *Fortune*, HBO spent more than $100 million a year on a 55-member development team based in Seattle,[38] and in return for its investment got a buggy system that crashed during the season finale of *True Detective* (in March of 2014), then again during the premiere of *Game of Thrones* (in April). In December of 2014, HBO pulled the plug on its internal development effort and outsourced its back-end delivery to an outside vendor: Major League Baseball Advanced Media.[39]

Outsourcing an important part of content delivery to another entertainment company might seem like a huge strategic risk, but in HBO's case it wasn't. Why? Because HBO and Major League Baseball compete in very different markets. But what if that wasn't the case? What if the very downstream technology partners that have become important to an entertainment company's media-distribution strategy decide to compete head to head with the company by producing their own content? Increasingly, that's what is happening in the entertainment industries. In the next chapter we'll take a look at the implications.

9 Moneyball

People ... operate with beliefs and biases. To the extent you can eliminate both and replace them with data, you gain a clear advantage.

John Henry, baseball team owner, quoted in Michael Lewis' book *Moneyball* (Norton, 2003)

We have the viewing data of everything.

Ted Sarandos, Chief Content Officer, Netflix

In the late 1990s, baseball scouts began looking at a young minor-league pitcher named Chad Bradford. An eccentric righty from rural Mississippi, Bradford had been racking up some impressive statistics, but he was an outlier and a curiosity: a "submariner" (that is, an under-hand pitcher) whose fastball averaged only 81–84 miles per hour. By major-league standards that was much too slow, and by *any* standards Bradford's delivery was bizarre. Here is how Michael Lewis, in his 2003 book *Moneyball*, described Bradford on the mound:

He jackknifes at the waist, like a jitterbug dancer lurching for his partner, his throwing hand swoops out toward the plate and down toward the earth. Less than an inch off the ground, way out where the dirt meets the infield grass, he rolls the ball off his fingertips. When subjected to slow-motion replay, as this motion often is, it looks less like pitching than feeding pigeons or shooting craps.[1]

The scouts just couldn't get themselves to trust a delivery like that. They admitted that Bradford's minor-league numbers were good but they *just knew* he would get eaten alive in the majors.

The thing was, if you compared Bradford's numbers with those of other pitchers who were getting called up to the majors, you had to think he deserved a chance. He almost never walked a batter. He averaged close to a strikeout per inning. Because his unusual delivery allowed him to release the ball much closer to the plate than an overhand pitcher could, his pitches took the same amount of time to reach batters as faster pitches from overhand pitchers—and that was confounding to batters, because reflexively they knew the ball wasn't moving as fast. Moreover, Bradford's pitches rose after he released the ball, then sank as the ball reached the plate, and thus the hits he gave up were almost always ground balls, not fly balls. As a result, he almost never gave up a home run, and rarely even gave up a double or a triple.

If you have read *Moneyball*, or seen the movie based on it, you know what happened next. One scout, an outlier himself, took an interest in Bradford. He didn't care what Bradford's pitching motion looked like. He just liked how Bradford could get results. He convinced the Chicago White Sox to give him a try. In 1998, the White Sox drafted him into their farm system.

Things got off to a less-than-promising start when the pitching coaches told Bradford bluntly that he was a "fringe prospect." But then Bradford went to work. Pitching for the White Sox' Triple-A team in Calgary, he dominated his opponents to such an extent that the White Sox felt compelled to call him up to the major-league team. They did, and he continued to shine. Working as a reliever, he retired the first seven batters he faced; he produced a run of twelve consecutive scoreless appearances during one stretch of the season; he gave up not a single home run; and he finished the season with an admirable earned run average of 3.23.

What followed was a storybook ending—overlooked pitcher finally wins the big contract he deserves—right?

Wrong. "The White Sox didn't trust Chad Bradford's success," Lewis writes. "The White Sox front office didn't trust his statistics. Unwilling to trust his statistics, they fell back on more subjective evaluation. Chad didn't look like a big leaguer. Chad didn't act like a big leaguer. Chad's success seemed sort of flukey."[2] So despite Bradford's performance that year, the team sent him back down to Triple A. He languished at that level until the end of the 2000 season, when Billy Beane, the general manager of the Oakland A's, scooped him up for the low price of $237,000 a year.

Beane had a plan. In charge of one of the financially poorest teams in baseball, he had decided—out of necessity but also conviction—to build a team made up of undervalued players. And the way to do that, he had decided, was to reject the game's traditionally subjective, instinct-based measures of success (what Lewis calls "the collective wisdom of old baseball men"[3]), which Beane believed had allowed a general fog of misjudgment and mismanagement to settle over major-league baseball. Instead, with the help of a motley crew of data geeks, he had begun to devise and employ analytically neutral metrics to determine value, and he was using those metrics to sign excellent players on the cheap. "By using baseball statistics," Lewis writes, describing Beane's philosophy, "you could see through a lot of base-ball nonsense."[4] To Beane, it seemed odd that no one had figured this out before. The traditionalists put a lot of emphasis on how many hits a pitcher gave up, for example. But did that make sense? Doesn't the number of hits a pitcher gave up depend to a significant degree on the performance of the fielders, rather than on the pitcher alone? If a shortstop shifts his position to face a batter and then lets a grounder through to the outfield, isn't *that* the reason the batter got a hit, not how the pitcher pitched? Why credit the pitcher differently for the very same pitch simply because the shortstop moved a little to the left or the right?

Beane had long felt that it was time for a new way of thinking. To find it, he had turned to Voros McCracken, an amateur statistician and baseball fanatic who had recently devised what he called the Defense Independent Pitching Statistic (DIPS), which, he argued, reflected a pitcher's abilities far more accurately than the number of hits batters got when facing him. Bradford's defense-dependent statistics were already good, but in the DIPS analysis they shot up, which is one of the reasons Beane decided he wanted Bradford on his team.

Beane and his data geeks had been applying this sort of analytical thinking not only to pitching but to every other aspect of the game for a few years, and in so doing had built a team of players that the rest of the league hadn't wanted. For a while no one had paid much attention to what they were doing—until the 2002 season, when the A's, despite having the second-lowest payroll in baseball, made it into the playoffs. How did they manage to do that? Moreover, how was it that during the previous three years, on average, they had paid about half a million dollars per win, whereas the big spenders in the league, many of whom hadn't made it to the playoffs, had paid more than six times as much?

The answer, of course, is that Beane and his team had tapped into the power of data. They had found new and better ways to measure players' value that stripped the old "gut feel" prejudices out of the analysis. At a time when most of the people running major-league baseball were former players who refused to believe that guys with computers could teach them anything about the sport to which they had dedicated their lives, Beane and his team had brought baseball into the data-driven age.

That brings us, in a roundabout way, to Netflix—another California organization that, in recent years, has used the power of data to locate hidden value in the market and to transform its industry.

• • •

We began this book with the story of how Netflix used the power of data in 2011 to license a television series that the traditional networks considered a dim prospect: *House of Cards*. Netflix decided to skip the supposedly essential step of producing a pilot episode and instead committed $100 million to producing two full seasons of the show—a move that was deemed crazy by most industry insiders at the time. But Netflix's data crunchers had analyzed the viewing patterns and preferences of the company's 33 million subscribers, and they felt confident that the show had a large potential audience. Soon enough, they were proved correct. *House of Cards* became a huge success.

But Netflix's story of industry transformation didn't begin with *House of Cards*. In 1997, Netflix had radically transformed the DVD-rental industry when, having detected what it felt was hidden value in the online DVD-rental market, it launched its delivery-by-mail service. In 2000, Netflix offered itself for sale to Blockbuster for $50 million, only to be turned down because executives at Blockbuster considered the online market insignificant. Blockbuster focused on protecting its in-store business and waited four years before launching an online competitor to Netflix. That delay proved costly. By 2010, Netflix had about 14 million customers, was the US Postal Service's fastest-growing client, and was mailing out hundreds of thousands of discs every day. That same year, Blockbuster filed for bankruptcy. "If [Blockbuster] had launched [an online subscription service] two years earlier, they would have killed us," observed Reed Hastings, the CEO of Netflix.[5]

Netflix, however, had little time to bask in its success, because in 2010 DVDs were on their way out and streaming was on its way in. Netflix embraced this change, cannibalizing its own DVD-rental business to launch an online streaming service that quickly became the biggest source of evening Internet traffic in North America. This flood of traffic contained a great amount of information, the company's executives recognized. In fact, it provided them with so much highly specific information about individual customers' habits and preferences that they could observe not only what movies and shows their customers liked but also how often they watched them, at what times they watched them, and even what parts of them they watched or replayed. No movie or TV studio had ever been able to tap into so much detailed information about its individual consumers—which is why, in 2011, none of them saw the potential of *House of Cards*.

To be clear, it isn't that the studios aren't trying to use data. They are. But they have run into a major obstacle when they have tried to implement "moneyball-style" approaches to decision making: the old guard. By this, we mean senior managers whose attitudes toward data-driven decision making are not unlike those that Michael Lewis attributed to the White Sox' front-office staff. ("You had general managers and

managers who had played the game," Lewis wrote. "How could someone who all they knew is computers tell them anything that would
make them more successful?")

The problem, in short, is cultural. "No one wanted to make decisions
based on data," one former member of a major movie studio's home-
entertainment group recently told us. "They don't know what to do
with it," another said. Moreover, these employees reported, the company's theatrical managing directors—in effect, the people who decide
which movies to make, and consider themselves tastemakers—har-
bored some profound cultural biases. They actively distanced themselves from movies that were unlikely ever to win awards or acclaim.
They also looked down on the home-entertainment business as a whole,
even though it makes up more than 50 percent of the profit on a typical
movie. "The big, sexy important part of movie production is the first-
run theatrical," one former staffer told us, and then proceeded to mimic
the theatrical staff's attitude toward the home-entertainment staff. "I
don't know how you live with yourselves just selling our old movies,"
he said. "You guys are the garbage men."

A senior executive at a major studio expressed nearly the same
cultural bias to us, though in different terms: "We have the creativity.
We're a content manufacturer. We *make* the content, which is a different business [than the distribution of home entertainment]. ... Frankly,
a certain subscription service [Netflix] got really lucky with the first
thing they launched." Maybe. But to us this sounds an awful lot like an
executive in the White Sox' front office telling the world that Chad
Bradford's success was just a fluke.

• • •

What we've been talking above, when you come right down to it, is a
clash between human expertise and data. Increasingly, the entertainment industries are competing against upstart content creators who
don't have long-standing biases against data-driven decision making.
For example, the executives at Amazon Studios—who, like their

counterparts at Netflix, are leveraging data gathered by a sophisticated online distribution network to create original films and TV shows—feel they have developed a much better way of making decisions about what content they produce. "We let the data drive what to put in front of customers," Bill Carr, Amazon's vice president for digital music and video, told the *Wall Street Journal* in 2013. "We don't have tastemakers deciding what our customers should read, listen to, and watch."[6]

And it isn't only in regard to movies that Amazon is thinking that way. Amazon hasn't been shy about expressing disdain for the way the publishing industry has clung to its old model. Amazon executives, Ken Auletta reported in *The New Yorker* in 2014, "considered publishing people 'antediluvian losers with rotary phones and inventory systems designed in 1968 and warehouses full of crap.'" That aptly captures the nature of the cultural shift we're talking about—one that publishing executives are waking up to. "I think we, as an industry, do a lot of talking," Madeline McIntosh, Random House's president of sales and operations, said in 2010. "We expect to have open dialogue. It's a culture of lunches. Amazon doesn't play in that culture. [It has] an incredible discipline of answering questions by looking at the math, looking at the numbers, looking at the data ... . That's a pretty big culture clash with the word-and-persuasion-driven lunch culture, the author-oriented culture."[7]

Even on the distribution side, the old guard—which includes independent booksellers—is having difficulty accepting what is happening. "We know our customers," Vivienne Jennings of Rainy Day Books ("Kansas City's Community Bookseller") told *The New Yorker* in 2014. "The other independents are the same. We know what [our customers] read better than any recommendation engine."[8] This may well be true at the local level, but in a globally interconnected marketplace localized expertise doesn't scale well—certainly not as well as data and algorithms. As one senior technology executive recently told us, it's "not even a fair fight."

Adjusting to a data-driven marketplace won't be easy for the creative industries. For 100 years, localized knowledge and expertise have been important sources of competitive advantage in these markets. That's because, historically, they have had very little first-hand information about consumers' behavior. Publishers are able to see aggregate statistics about the number of sales on any particular title, but they have very little information about the individuals purchasing a title. Similarly, music labels can buy Arbitron ratings that estimate how many people heard their songs on the radio in a certain time frame, but they have almost no information about those listeners as individuals—who they are, what they enjoy about the content, and what other content they enjoy; and motion-picture studios don't interact directly with their customers in any of their distribution channels (theaters, home entertainment, or television broadcast)—a point that was made forcefully by Sony's CEO, Michael Lynton, in 2014. "We don't have that direct interface with the American public," he said in response to a question about Sony's ability to release the film *The Interview* after the November 2014 cyber-attack on the studio. "We need to go through an intermediary to do that."[9]

In the absence of detailed customer data, the creative industries have made decisions about what content to produce by combining aggregate statistics (such as Arbitron ratings and Nielsen statistics) with data from small samples (such as focus groups) and "gut feel" on the part of individuals within the industries who understand the market. Because "gut feel" is so important to the organization, the people who are effective at evaluating talent are the ones who typically rise into positions of power in their firms. The result of all this is that most entertainment firms have relatively little institutional strength or political capital associated with data analytics, particularly compared to the institutional capital they have invested in "gut feel" decision making. "Data can only tell you what people have liked before, not what they don't know they are going to like in the future," John Landgraf, the president and general manager of FX Networks, told the *New York Times* in 2013.[10] "A good high-end programmer's job is to find the white spaces in our collective psyche that aren't filled by an existing television show," he said. And that, he continued, happens only "in a black box that data can never penetrate." The majors have thought this way for a long time, and they have enjoyed great success. If that strategy hasn't been problematic before, why should it be problematic now? To answer that question, let's return to Billy Beane and the Oakland A's.

• • •

We have talked a lot about the competitive success of the Oakland A's in adopting "moneyball-style" decision making, but their success with that strategy was short-lived. Although Billy Beane's innovations had a substantial effect across Major League Baseball in terms of how personnel decisions are made, they had very little effect on his team's long-term competitive advantage. Sure, "moneyball" techniques may have given the A's an edge for a year or two, but the other teams soon figured out how to copy the A's strategy and restored the sort of competitive "parity" that makes it nearly impossible to win championships without spending a lot of money.[11] One might conclude from this example that we will see a similar evolution in the entertainment industries. Netflix, Amazon, and Google may gain short-term advantage from innovative ways of using data, but the majors will find it easy to copy those techniques and will maintain their powerful position in the market. We disagree. We believe it is going to be much harder for the entertainment majors to catch up to their new rivals, mainly for two reasons.

The first reason is cultural. Major-league baseball teams were all very similar when it came to how the organizations used data. Thus, when they needed to change their cultures to adopt a new managerial style, they were all starting from the same place. But this isn't true when it comes to the new technology companies in the entertainment industries. As we discussed previously, when it comes to using data, the cultures in technology firms are very different from those in the entertainment industries. This was driven home to us in February of 2009 when Richard Hilleman, the Chief Creative Officer at EA Games, came to Carnegie Mellon to talk with our class about technological change in the gaming industry. During the discussion, someone asked him why the publishing, music, and motion-picture industries were having so much trouble adopting data-driven decision making. "You have to remember," Hilleman said, "that decisions in these industries have always been made based on someone's 'gut feel' about what will sell in the market, and the people with good gut instincts are the ones who rise to positions of power in their companies. The problem is, these companies are now competing against companies like Google, Amazon, and Apple who don't make gut feel decisions—they make quantitative decisions based on what their data tells them."

The second reason has to do with access to data. In major-league baseball, everyone had access to the same data. Any team could go to Stats Inc. or Elias Sports Bureau and buy the same data Oakland was using to make decisions, and then use those data to replicate Oakland's analysis and decision-making techniques. But access to data is very different in the entertainment industries than in baseball.

Think for a minute about the volume of data the new online distribution platforms gather and control. Netflix, as we have pointed out, knows what content each of its consumers watches, when they watch it, on what device, what scenes they skip, and what scenes they watch over and over again. Likewise, Amazon gathers great amounts of data from the customers of its streaming service, which it then can combine with the purchase histories and searches of those customers in other e-commerce categories. The YouTube platform provides Google with similar data that it can link to a customer's behavior on Google's other platforms.

And this exchange of information isn't only from customer to platform. Once platforms have learned customers' preferences, they can market products directly to those customers on the basis of what they have learned. They can recommend specific products to customers on the basis of their observed behavior, they can directly measure the effectiveness of different promotional strategies, they can design specific promotional campaigns for different types of customers; they can even use data to target new customers, or attempt to reinvigorate their relationships with longtime customers who have reduced their purchases. This two-way process can create a virtuous cycle in which information that customers provide creates a better experience for those customers, who then develop greater loyalty and use the platform more, thus revealing more about their preferences to the firm.

Of course, in net terms this could be positive for content creators if the platform companies shared their data and allowed creators to market directly to the platforms' individual customers. But they don't. As is summarized in table 9.1, the platform companies share almost no customer-level data with their upstream "partners." For example, Apple's sales reports provide its suppliers with transaction-level data on what consumers purchase on Apple's platform, along with an ID number and a ZIP code for each customer. Of course, having an individual customer ID is a good start, and we have talked to several of the major entertainment firms about using this data to create innovative marketing strategies that target specific customers with specific promotions. But only Apple knows which customer corresponds to which ID number, and without Apple's cooperation the majors have no way to reach those customers. Also, whereas suppliers only see purchases of their own content, Apple sees purchases across all suppliers and all content sold on the iTunes platform—movies, television shows, and music.

But at least Apple shares *some* customer-level data with its partners. Amazon, Google, and Netflix have adopted an even more extreme approach. Their sales reports provide *no* data on customers to their

Table 9.1
Data sharing between online platforms and content owners.

Distributor	Transaction-level data	Customer-level data	Direct promotion to customers
iTunes	Yes	Limited (ID and ZIP code)	No
Amazon	Yes	No	No
Google Play	No	No	No
Netflix	No	No	No

industry partners. Until the mid 2000s Amazon shared the ZIP codes from which sales had originated, but recently Amazon removed even that minimal information from its reports. Google Play and Netflix don't even provide transaction-level data to their industry partners. They only share the aggregate number of sales or views of a studio's content within a specific market.[12] For example, one content creator we spoke to receives quarterly reports from Netflix documenting the number of views of their content aggregated across all of Latin America. They received no data specific to, say, Mexico or Brazil.

Why are the platform companies so stingy about sharing customer data? After all, data sharing is common in brick-and-mortar retail markets. Grocery stores and other physical retailers regularly share detailed customer-level data with upstream manufacturers such as Procter and Gamble, Coca-Cola, and Pepsico, and allow those manufacturers to promote products directly to the retailer's customers. In fact, in 1998 when Amazon launched its online video store, it used the promise of access to detailed data on customers to entice studios to make their videos available for sale on Amazon's website. According to Jason Kilar, who headed the effort, the studios uniformly resisted working with Amazon's video store until they were won over by what Amazon said its data could do for them. "We had to beg for meetings back then," said Anne Hurley, the founding editor of Amazon's DVD page. "What helped sell the studios was the tech edge Amazon offered. We could share our search results. We could tell them what customers really wanted—information they had never been able to get their hands on before. So studios could focus their attention on releasing those titles that already had a built-in buying audience."[13]

But Amazon isn't as forthcoming with its data today. Why not? Well, for one thing, Amazon is no longer a hungry Seattle startup; it's a retailing behemoth with enough market power to dictate terms. With detailed data and a lot of data scientists, Amazon can monetize its strategic assets and use them as leverage in negotiations with its partners.

If you want to promote your content on Amazon's site or to Amazon's customers, you had better be ready to pay.

More important, Amazon's entry into the content-production business has made the other content producers look less like partners and more like competitors. In 1998, when Amazon was merely a prospective distributor of motion-picture content, helping studios discover what their DVD customers really wanted made good business sense. Less so today. Amazon spent more than $100 million on original series in the third quarter of 2014 alone, and in early 2015 it announced plans to produce about 12 movies per year, with production budgets ranging from $5 million to $25 million.[14] Google's YouTube channel is also getting into original production. It has set up studio space for its content creators in Los Angeles and five other major cities, and plans to release at least ten of its own original movies and series on its new subscription service, YouTube Red, in 2016.[15] But both Amazon and Google are playing catch up to Netflix. In 2015, Netflix tripled its output of original programming, with more than two-dozen new series and 320 hours of new content, which by some measures surpasses the output of HBO and FX, both longtime leaders in the production of original content for cable audiences.[16] And Netflix isn't slowing down: The streaming giant announced plans to produce 600 hours of original content in 2016.

By keeping proprietary control over their data on customers, the big platform companies are able to use their data to evaluate the potential market for original content, and use their direct connections with customers to do highly targeted, preference-based marketing—something that can't be done with Nielsen estimates and focus-group data. "The real advantage we have," the Netflix spokesman Jonathan Friedland has said, "is not in picking the perfect content, it is in marketing it more efficiently."[17] In the case of *House of Cards*, as we pointed out in our opening chapter, this meant targeting customers by developing nine separate "trailers," some of them focusing on Kevin Spacey (for those who had watched his other acting work), some featuring David Fincher's directing style (for those who had watched his other directing work), and some focusing on the show's female characters (for those who were fans of movies with strong female leads).[18]

In short, when it comes to producing original content, the new downstream platforms possess three important advantages.

First, as we noted above, their hoards of proprietary data and their culture of data-driven decision making allow the new downstream platforms to identify and produce "blockbuster" content that the traditional producers have overlooked while using their "gut feel" approach to decision making.

Second, because of the on-demand nature of their platforms and their ability to promote content directly to individual consumers, the new downstream platforms can profitably produce "long-tail" content that wouldn't be profitable in traditional channels. In their mass-market channels, movie and TV studios have to focus on shows with mass appeal, whereas the downstream platforms don't. Roy Price, the head of Amazon Studios, summarized this approach in an interview with the *Hollywood Reporter*: "Let's say you had a show where 80 percent of the people you show it to think it's pretty good. They might watch it, but none of those people think it's a great show, nor is it their favorite show. But then you have another show where only 30 percent of people like it. For every single one of them, they're going to watch every single episode and they love it. Well, in an on-demand world, show No. 2 is more valuable. That really changes how you approach it, because what you need to do is get more specific. It's less about following generic, general rules for creating television and more about finding a specific voice and a specific artist that people are going to be a fan of."[19] This characteristic of on-demand programming may explain why Netflix was willing to acquire the rights to produce additional seasons of the show *Arrested Development* after it was canceled on broadcast television. The show had a distinct "voice" and a loyal fan base, but it wasn't a big enough hit to be profitable in a broadcast world. It also might explain why Netflix was willing to sign Adam Sandler to a four-movie deal, and why Amazon was willing to sign Woody Allen to make a television show. Both actors are acquired tastes, but both have strong fan bases that are hard to reach without the detailed data and customer connections that Amazon and Netflix possess.

Third, downstream platforms can create a strong connection between their content and their platform brand. This opens up new ways to increase customer loyalty and new cross-promotion options. Moreover, it will be difficult for the major entertainment companies to replicate these brand-loyalty and cross-promotional opportunities. Because the majors have never needed to develop a strong tie between their content and their brand, almost no one outside the industry knows (or cares) what studio produced *Jurassic World*, what label produced Taylor Swift's last album, or what publisher put out *The Da Vinci Code*.

So far in this discussion, most of the stories we have told have had to do with the motion-picture business. But "big data" is becoming increasingly important in the music and publishing industries too. Pandora (through its Music Genome Project), Shazam, and Spotify have now gathered detailed data about their customers' preferences that will prove useful in the marketing of new artists. In fact, Spotify's CEO, Daniel Ek, argues that the data his company has gathered have already given it an enormous competitive advantage: "We've been doing this for years, and what we've built is the largest set of data of the most engaged music consumers."[20] Likewise, Shazam, which allows a user to identify music he or she has heard anywhere by means of a smartphone app, has developed a competitive advantage because it documents the individual details of 20 million searches every day. The predictive power of Shazam's data has made its app hugely popular among music agents around the country, and in February of 2014 the company announced that it would use its data to produce music for a new Warner Music Group imprint.[21]

In the end, these firms derive much of their competitive advantage from the power of data to generate customer loyalty and market power, and then to facilitate vertical integration into content production. When Amazon went into the book business, few people recognized that bookselling wasn't the company's primary focus. "I thought he was just a bookstore, stupid me," John Sargent, the CEO of Macmillan, said in 2011 about what Jeff Bezos was really up to with Amazon. But books, it turned out, were just Bezos' way to get the data. Books, Sargent says he eventually recognized, were Amazon's "customer-acquisition strategy."[22]

• • •

When we talk to entertainment-industry executives about the challenges posed by the new data "insurgents," we typically hear the following four responses, sometimes in exactly this order:

• You can't use data to make creative decisions. If you do, you'll interfere with the creative process and destroy the business.
• We have our own data, and for years we have been using data to make decisions. The data that these new companies are using aren't much different from the data we have always used.
• These companies rely on us for content. If they get too powerful, we'll just stop licensing our content to them.
• All we have to do is open our own streaming channels, and we can get all the customer-level data we want.

Let's examine these arguments in turn.

You can't use data to make creative decisions. If you do, you'll interfere with the creative process and destroy the business.

We believe there are two flaws in this logic. The first is in the premise that Netflix is using data to interfere with the creative process. "We don't use data to influence creative at all," Ted Sarandos told the National Association of Television Program Executives in 2015. "Our data is mostly used to say 'Wow, there's a real *there* there for this show. All the elements are there for this to be a great big show, and therefore you invest heavily.'"[23] An analogy to "moneyball" works nicely here. The Oakland A's weren't using their data to tell Chad Bradford how to pitch; they were using it to evaluate how effective his pitching style was.

The second flaw in the argument is that creators seem to be getting more, not less, freedom from the new data-driven entrants. Kevin

Spacey made this point eloquently in his keynote address at Content Marketing World, in 2014. His remarks are worth quoting at length:

> In recent years we have seen an explosion of compelling, dynamic programming with characters and narratives that were nuanced and full of texture. I'm referring to shows like *The Sopranos*, and *Weeds*, and *Homeland*, and *Dexter*, and *Six Feet Under*, *Deadwood*, *Damages*, *Sons of Anarchy*, *Oz*, *The Wire*, *True Blood*, *Boardwalk Empire*, *Mad Men*, *Game of Thrones*, *Breaking Bad*, even *House of Cards*.
>
> And these programs, frankly, would not have been made before 15 years ago. Because before then, most network executives thought all characters needed to be really nice, and be really good at their jobs, and to be good family people. Barney Fife wasn't a bi-polar CIA officer who falls in love with his target. Mary Tyler Moore wasn't a high school chemistry teacher with a proclivity for cooking meth. And I think this evolution in programming—what I believe represents the third Golden Age of television—comes squarely down to the fact that the creatives have more control over the story than ever before. Because in the old days the power was concentrated in the hands of a few—the studios, the networks, the executives. Those were the ones that sat around making decisions about what would get made, how it would get made, and who would get to see it.
>
> I mean, I even remember in my old first early jobs in television, I would see them, hovering around the cameras in suits. The network people. (I see network people.) Sticking their fingers in every creative decision, having opinions about everything, asking why my hair was that way, why I was wearing that tie, why I was acting that way. It left a bad taste in my mouth, and it motivated me to focus on film and theater. But making *House of Cards*, on the other hand, with Netflix has been the antithesis of my first experiences in television. In fact, it's been the most fun, and the most creatively rewarding experience I've ever had in front of the camera.[24]

Data-driven content, it should be noted, is also winning awards for its creativity. At the 2015 Golden Globe Awards, for example, Amazon won the Best Comedy award with *Transparent*. What did it beat out? *Orange Is the New Black* (Netflix), *Silicon Valley* and *Girls* (HBO), and *Jane the Virgin* (CW). And in 2016, Netflix received more Golden Globe nominations (eight) than any other television network, ending HBO's 14-year streak as the network with the most nominations.[25] In fact, the number of nominations Netflix received in 2016 fell just short of the *combined* number of nominations received by the traditional broadcast networks: ABC (four), Fox (four), CBS (two), and NBC (zero).[26]

Another form of recognition comes from the number of established artists who are choosing to work with these new data-driven producers rather than with the traditional studios. This has caused some in the industry to express concern about a "talent drain" in which actors, writers, and other creative professionals are beginning to take their talents to the new platforms.[27]

We have our own data, and for years we have been using data to make decisions. The data that these new companies are using aren't much different from the data we have always used.

Yes, the majors have been using data to make decisions for years—but very little of the data is proprietary. Anyone in the industry can buy audience estimates from Nielsen or Arbitron, sales estimates from Rentrak, SoundScan, or BookScan, or online audience statistics ComScore. Moreover, the data that Netflix, Amazon, Apple, and Google collect are much more detailed than existing industry statistics and much more comprehensive than what can be learned from focus groups. Perhaps most important, the data the platform companies collect enable direct promotion and direct customer engagement in ways that aren't possible with Nielsen audience estimates or Bookscan sales estimates.

These companies rely on us for content. If they get too powerful, we'll just stop licensing our content to them.

Executives in the entertainment industries often argue that, if push comes to shove, they will remove their content from the new platforms, which will put the platforms out of business or at least greatly reduce their market power.

As a stand-alone strategy, this strikes us as misguided. As NBC discovered in its fight with iTunes, in the absence of an established legal alternative, removing content from these platforms may simply shift customer demand toward digital piracy. Even with a legal alternative, Netflix, Amazon, Google, and others may already be too powerful to be shut out of the game altogether. Studios, labels, and publishers rely on the revenue these platforms provide,[28] and reap important spillover benefits by allowing consumers to use digital platforms to discover their content.

All we have to do is open our own streaming channels, and we can get all the customer-level data we want.

This strikes us as a good start, but one that, by itself, is still incomplete when it comes to competing against new platform services. Online consumers place a high value on convenience, and, as we mentioned earlier in this chapter, most content isn't branded under the producer's name. Thus, if Fox were to open an online streaming channel for movies, or even for television shows, few consumers would know which of their favorite programs they could find there. And even if they did know where to find their content across a number of producer websites, consumers would much rather get access to all their content in one place than have to learn the intricacies of doing so for content on multiple studio websites. Centralization also generates benefits for producers. Even if an individual firm opens its own streaming platform with its own content, and collects detailed viewer behavior, those data aren't anywhere near as valuable as what Netflix or Amazon can see: viewers' behavior across all the content available through its service.

• • •

Let's take stock of where we are. In chapters 1–4 we talked about the economic foundations of the entertainment industries. Historically, as we laid out in detail in chapter 2, the entertainment industries have exhibited strong economies of scale and strong barriers to entry in the production of books, music, and motion pictures. Because of these economic characteristics, a small number of powerful publishers, labels, and studios have been able to exercise a great deal of control over downstream promotion and distribution channels and upstream artists. In chapter 3, we discussed how firms also rely on the ability to control when and how their customers can gain access to content to sell their products profitably.

In part II of the book we have argued that a perfect storm of technological change is altering the sources of power and profit in the entertainment industries. The viability of long-tail markets, the widespread availability of digital piracy, a host of new production and distribution options for artists, the rise of powerful downstream distributors, and the ability of these distributors to control detailed data have all figured in shifting power away from the firms that control content and toward the firms that control the customers.

As we have argued in this chapter, we believe that the major studios, labels, and publishers face two main challenges in adapting to these new competitive realities: (1) organizational biases toward protecting existing business models and preserving the existing "gut feel" approaches to decision making and (2) a lack of access to the valuable customer-level data that have become important to the production and promotion of entertainment. We believe that the majors will have to address these two issues to continue to thrive in the market for entertainment. In part III we will discuss the strategies we believe the majors can use to do so, beginning with organizational change.

III A New Hope

Any attack made by the Rebels against this station would be a useless gesture, no matter what technical data they have obtained.

Admiral Motti, in *Star Wars Episode IV: A New Hope*

10 Pride and Prejudice

I came to you without a doubt of my reception. You showed me how insufficient were all my pretensions.

Jane Austen, *Pride and Prejudice*

I am purely empirical. I am not attached to any romantic notion of how this business should be run. I am only driven where the evidence takes me.

Gary Loveman, CEO of Harrah's Entertainment[1]

If the entertainment industries hope to prosper in the rapidly changing business landscape of the digital age, they will have to harness the power of detailed customer-level data and embrace a culture of data-driven decision making. To do that, they will need to make significant organizational changes—no easy feat in industries that hardened structurally decades ago, long before customer-level data emerged as a useful tool. Making such changes will not be easy, logistically or culturally, but it will be necessary to compete with the data-driven entrants we discussed in the preceding chapter. Nothing illustrates the benefits of organizing around data better than the story of how Harrah's (now Caesars) Entertainment gained an edge over its competitors and became a market leader in the gaming industry during the early 2000s.

• • •

Harrah's was a twentieth-century success story.[2] The company's founder, William Fish Harrah, moved to Reno in 1937 and soon set up a small bingo parlor, and then a casino. The betting parlors in Reno at the time were saloon-style operations, dark and rough and dingy. Harrah thought he would be able to attract lots of customers by creating something different: a clean, brightly lit, luxurious venue that would suggest not sinful escape but good clean fun.

People flocked to Harrah's casino. In the decades that followed, he built more like it elsewhere and became, as the Harvard Business School professor Rajiv Lal put it, "the man who industrialized gambling."[3] In 1955, using Harrah's as his company name, Harrah built the world's largest stand-alone gambling venue, on the shores of Lake Tahoe; its 850-seat theater-restaurant regularly brought it top entertainers from all over the country. Increasingly, Harrah's casinos became destinations. Harrah built grand hotels alongside them. He died in 1978, but the company continued to grow. In the 1970s and the 1980s many states legalized gambling, and in the 1990s Harrah's launched an aggressive expansion strategy that made it the first nationwide casino business. "By 2000," Lal writes, "Harrah's Entertainment, Inc. was well known in the gaming industry and operated casinos in more markets than any other casino company. ... Harrah's operated land-based, dockside, riverboat, and Indian casino facilities in all of the traditional and most of the new US casino entertainment jurisdictions."[4]

Harrah's was flying high in 2000, but the landscape underneath it was changing. In the US, no other states had legalized gambling, and the company's expansion strategy was no longer viable. The market was now limited, and Harrah's found itself having to compete with flashy new companies that were attracting large numbers of customers by building fantasyland resorts similar to what had been pioneered in Las Vegas at the Mirage (shark tank, wild animals, fake lava-spewing volcano) and the Luxor (monumental glass pyramid, reproductions of Egyptian temples, statues of pharaohs). In Las Vegas and elsewhere, these companies surrounded their casinos with glitzy shopping malls, fancy restaurants, luxurious spas, and all manner of over-the-top entertainment in a concerted effort to appeal to customers other than gamblers. And it was working. In 2001, consumers in Las Vegas spent three times as much money on shopping, dining, and entertainment as they did on gambling. And even though the gambling market had expanded widely in the previous decades, at the end of the 1990s Nevada and Atlantic City still took in 40 percent of the $31 billion made annually in gambling in the United States.

Over the course of 50 years, Harrah's had very successfully built a profitable national network of casinos, each of which operated largely autonomously in its own market. But almost all of Harrah's revenue came from its casinos, not from its stores, restaurants, and entertainment offerings. The company didn't have the resources to compete on the new terms. Reinventing itself overnight, all over the country, as a builder of monumental casino resorts wasn't a realistic option. If Harrah's wanted to stay competitive, its CEO, Philip Satre recognized, it would have to find a different way. In the mid 1990s, Satre committed the company to a plan. "Customer loyalty was really our competency," he would later recall, "and we decided that we could become an industry leader based on that skill."[5] To that end, in 1997 Satre launched Total Gold, a loyalty program modeled on airlines' frequent-flyer programs. Customers at a casino would earn credits on their cards as they gambled, which could be put toward the standard sorts of awards offered in the industry: free meals, free hotel stays, free show tickets. But because Harrah's casinos were autonomous operations, the loyalty cards were valid only at the Harrah's property that had issued the card, and each property determined its own marketing programs for its customers.

Satre soon recognized how much more valuable a national loyalty program would be—one that would allow customers to earn credits that could be used at any Harrah's casino—and began investing in that idea. At about the same time, in 1998, Satre hired Gary Loveman, a professor at the Harvard Business School, as his new chief operating officer, and gave him a clear mandate. "When Satre hired me as COO," Loveman wrote in 2003, "he said he wanted to change Harrah's from an operations-driven company that viewed each casino as a stand-alone business into a marketing-driven company that built customer loyalty to all Harrah's properties."[6]

When Loveman arrived, he immediately recognized that the Total Gold program was inadequate because it varied from casino to casino and gave customers no incentive to make Harrah's their gambling destination all over the country. At the same time, he recognized that, despite its obvious flaws, the program offered a way forward for Harrah's: the sophisticated mining and analysis of data. "While Total Gold wasn't much good for keeping customers loyal to Harrah's," Loveman wrote, "it was quietly digging our future diamond mine."[7]

Building the use of data into Harrah's business model wasn't going to be easy, Loveman recognized. That's because the organization as a whole hadn't evolved to share information among casinos. "Each property was like a fiefdom," Loveman wrote, recalling the situation he encountered when he started, "managed by feudal lords with occasional interruptions from the king or the queen who passed through town. Each property had its own P&L and its own resource stream, and the notion that you would take a customer and encourage them to do their gaming at other properties was not common practice."[8] This structure wasn't arbitrary; it came from a desire to give property managers incentives to improve their local operations by forcing them to compete for resources with other properties. But this silo-like organizational structure was incompatible with Loveman's vision of company-wide data-driven management.

Loveman decided that his first priority had to be organizational change. He began, with Satre's support, by requiring the casino managers and division presidents, who had previously reported directly to the CEO, to report to Loveman instead. This was a signal, as Loveman put it, "that customers belonged to Harrah's and not simply to one of its casinos." This wasn't a trivial change, or one that casino managers took lightly. These managers were, for the most part, industry insiders who had worked their way up in the business from the bottom. The autonomy and power of running their own casino was their reward, and having their marketing programs dictated from headquarters by an industry outsider was a direct challenge to their power and control. The change also represented a potential challenge to their income. Their bonuses and other incentives were based on revenue generated by their casino, and the idea of giving up data to encourage customers to take their gaming to another casino—even another Harrah's casino—was seen as a threat.

Not all of Harrah's employees were able to make the switch to the new reporting structure and the reduction in their power and autonomy. Loveman replaced resistant general managers at important Harrah's properties in Reno and Las Vegas shortly after he arrived.[9] After centralizing customer marketing, Harrah's lost a quarter of its casino hosts, the employees who previously had the power and authority to determine what complementary incentives, or "comps," guests at their casinos would receive.[10]

Although painful, changing the organization's structure allowed Harrah's to centralize its data in a sophisticated network that linked all of its properties, and to focus methodically, for the first time, on extracting value from that data. "By tracking millions of individual transactions," Loveman wrote, "the information-technology systems that underlie the program had assembled a vast amount of data on customer preferences. At the core of the Total Gold rewards program ... was a 300-gigabyte transactional database that recorded customer activity at various points of sale—slot machines, restaurants, and other retail areas in our properties. Database managers fed that information into our enterprise data warehouse, which contained not only millions of transactional data points about customers (such as names, addresses, ages, genders) but also details about their gambling and spending preferences. The database was a very rich repository of customer information."[11]

In addition to changing reporting structures and making data analytics a "C-level" function, Loveman signaled the new organizational importance of data analytics by hiring as members of his own senior staff people with strong quantitative backgrounds. This team, which Loveman referred to as "propeller heads," included two powerful senior

vice presidents in charge of customer relationships and customer loyalty: Richard Mirman, a former University of Chicago mathematician,
and David Norton, a former analytics executive with American Express,
Household International, and MBNA America.

Loveman infused data analytics into the organizational culture by
insisting that all decisions be based on rigorous analysis and testing,
not on hunches about what might work. "When I meet with our marketers to discuss anything that we've done that is new," Loveman said,
"I ask, 'Did we test it first?' And if I find out that we just whole-hog
went after something without testing it, I'll kill 'em. No matter how
clever they think it is, we test it."[12]

Using their newly integrated data platform and analytics focus,
Loveman and his team began to learn all sorts of surprising things.
For example, they were able to show property managers that Harrah's
new focus on centralized customer management wasn't cannibalizing
local revenue. "We challenged the premise that a trip to Las Vegas
would take away a trip to Tunica, Missouri," Mirman said. "We did
tests, and we showed [property managers] that that wasn't the case.
Once we built the systems, we were able to prove that there is a lot of
cross-market play."[13]

They also found that 26 percent of their customers generated
82 percent of Harrah's revenue. But more important, when they looked
at those customers individually they learned that their most profitable
customers weren't the traditional high-rolling "whales" that everyone
else in the industry was focusing on. Instead, they were middle-aged
adults and senior citizens who enjoyed playing the slot machines.
With this knowledge, Loveman designed a strategy to go after "low
rollers" in a business in which everyone else was going after "high
rollers."

With customer-level data, Loveman and his team were also able to
predict the lifetime value of a new customer on the basis of very little
information about what games the person played, how much he or she
bet, and how quickly he or she played. Specifically, they could take the
information they observed about a customer and generate a prediction

about how valuable that customer would be in the future. They could then compare that prediction against what they actually observed about a consumer's play at Harrah's and determine what promotions they should send to that person. For example, a consumer who the model said was a heavy gambler, but who Harrah's didn't see much in Harrah's casinos, probably was spending a lot of time at other casinos, and Harrah's could target him with a promotional program designed to increase his loyalty to Harrah's. Likewise, if Harrah's observed that a regular customer was visiting less frequently, it could target that customer with a retention effort.

The new integrated data platform also allowed Harrah's to design and run experiments to see which marketing strategies were most effective. "We run everything at Harrah's with control groups," Loveman said. "It's like, 'You don't harass women, you don't steal, and you've got to have a control group.' This is one of the things that you can lose your job for at Harrah's—not running a control group."[14] This, too, turned up some surprises. In one experiment, the company offered some customers a typical loyalty reward of a free room, two steak dinners, and $30 worth of free chips (total value $125), and offered another group only $60 worth of free chips. Against expectations, the latter promotion turned out to be twice as profitable as the former. Harrah's also ran experiments that analyzed how customers chose which slot machines to play, with a level of detail that even included the color of the machine's background. With this information gained, Harrah's was able to change the design and layout of its slot machines to better match customers' preferences.

In short, Harrah's marketing became highly quantitative. Instead of basing its marketing strategies on observed play, the company chose to focus on *predicted* play, deduced from a relatively small set of observations. This difference is critically important. A casino that considers only observed play will assign low value to an infrequent customer, whereas a casino studying predicted play may be able to discern that an infrequent customer at Harrah's is a regular gambler elsewhere and therefore should be considered a high-worth customer.

This quantitative approach allows a strong emphasis on personalization, which is what Harrah's recognized would best generate customer loyalty. As Richard Mirman put it, the company wanted a customer to think "I want to go to Harrah's because they know me and they reward me like they know me, and if I went somewhere else, they would not."[15] Harrah's also discovered a virtuous cycle in the use of detailed data and proprietary algorithms to gain customers' loyalty. "The farther we get ahead and the more tests we run, the more we learn," Loveman is quoted as having said. "The more we understand our customers, the more substantial are the switching costs that we put into place, and the farther ahead we are of our competitors' efforts. That is why we are running as fast as we can."

Harrah's transition to data-driven management involved a strategy based on three main principles:

• centralizing data across the company and reorganizing around data analytics as a function with c-level importance and authority
• insisting that all decisions should be informed by data, ideally from controlled experiments
• treating customers as individuals and designing marketing strategies on the basis of individual behavior.

The strategy worked. By 2003, Harrah's had posted sixteen straight quarters of revenue growth, and in 2002 it earned $4 billion in revenue and $235 million in net income.[16] Gary Loveman was understandably proud of his success in reorganizing the company to take advantage of the power of data. "We've come out on top in the casino wars," he wrote, "by mining our customer data deeply, running marketing experiments, and using the results to develop and implement finely tuned marketing and service-delivery strategies that keep our customers coming back."[17] Loveman replaced Satre as Harrah's CEO in 2003 and held that post until he stepped down in 2015. During that time he transformed Harrah's into the largest gaming corporation in the world, acquiring Caesars Entertainment Inc. and expanding Harrah's from 15 casinos in 2003 to more than 54 in 2013.[18] But his most important acquisition was data on customers. When he stepped down as CEO, his loyalty program had 45 million members and was estimated to be worth more than $1 billion.[19]

• • •

We believe the story of Harrah's has important lessons to teach the entertainment industries, which will have to implement a similar strategy to meet the needs of a new data-rich marketplace. In the remainder of this chapter, and in the next chapter, we will discuss these changes. For simplicity, we will focus on the motion-picture business, but with an understanding that the changes we're proposing, although focused on the distinctive aspects of the motion-picture industry, are similar to the changes that will be necessary in the other entertainment industries.

Let's begin with "data silos." Before Gary Loveman's arrival, Harrah's data on customers were spread throughout the company and owned by the local casinos. A similar situation prevails in motion-picture studios today. Data typically are spread throughout the company and owned by individual business units (theatrical, television, home entertainment, and so on). These units, just like the old Harrah's casinos, are often unwilling to share their data with one another, because they're worried about losing a source of competitive advantage within the company. An employee at a major studio described the business units at her company to us as made up of about forty "fiefdoms," each with its own president. "They all want to do their own thing," she noted, "and the overall health of the company gets lost." This sort of structure may have made sense in the pre-data age (for example, as a managerial tactic to encourage competition between business units), but it makes no sense now that companies increasingly need to tap into the power of data to thrive in a data-driven marketplace. Harrah's recognized this early and gained an advantage over its competitors by centralizing data analytics and putting them into the hands of a new C-level team. We believe studios should follow a similar path, and we see four main ways in which making data analytics a centralized C-level function might help the studios and other major entertainment companies compete with Google, Amazon, Netflix, and other new entrants.

The first is obvious but worth stating for the record. Data are most useful when linked across datasets and seen as a whole. This is particularly true of customer data and market data, which, when "siloed," represent a valuable trapped resource that, if linked together, could be deployed to great competitive advantage. Consider the staggered way most movies are released: first in theaters, then on DVD, then via various TV and Internet outlets. With this approach, the pricing and marketing decisions made in one release window naturally affect sales in the other windows—but at present the studios don't insist that the managers of these different channels share their data or centralize their decision-making. To succeed going forward, they're going to have to.

Second, a centralized structure will increase the effectiveness of analytics talent within the company. Analytics tasks require a variety of skill sets (experimental design, statistical inference, econometric modeling), and it's rare to find these in a single person. Centralizing the function would make it much easier for analysts with different training and skills to collaborate. And it would also allow for the centralized collection of data, which would prevent redundancies across business units.

Third, raising data analytics to a prominent position in the organization will make retaining existing talent and attracting new talent easier. It will make clear to everyone in the company that data analytics is a priority, and will provide a career path for individuals with strong analytic skills.

The fourth reason is the most important: Centralizing the reporting structure for analytics will help the company as a whole get objective answers to controversial questions. When analytics talent is hired by the individual business units, analysts can be pressured to make the data say what their bosses want the data to say. We've heard stories of powerful executives' becoming enraged upon seeing their analysts' results, not because it's inaccurate, but because it doesn't confirm the executive's earlier decisions, or because the result might hurt the business unit's short-term bottom line. The risk here is obvious. If analysts believe that their employment is dependent upon producing a certain answer, they will be tempted to ignore the evidence and to pursue a pre-determined result. In the long term, no one wins if company executives are basing their decisions on flawed analysis.

Of course, moving analysts too far away from the business units can create other problems. If the analysts aren't familiar with the business units' operations, they won't be able to identify or reliably answer the business units' most pressing questions. And, if the business units don't trust the analysts' knowledge, the business units won't use them in the first place. So how can analysts maintain their objectivity while retaining a deep knowledge of the business unit's unique needs? We spoke to a senior executive at one of the major platform companies, who told us that in his company the analysts are hired by and report to a centralized team of data scientists. The team meets regularly so that the analysts can discuss their projects and draw on the different skill sets of their teammates. However, the analysts' offices are located within the individual business units, which allows them to gain a detailed understanding of the business unit's needs.

How might a data-driven culture look in the entertainment industries? To answer that question, let's begin by asking how data analytics and experimentation might help industry executives develop new approaches to the classic "four P's" of marketing: What *product* should I sell, what *place* should I sell it in, what *price* should I set for it, and how should I *promote* it?

Product

After content has been created, data can play a powerful role in deciding the most effective way to bring it to the market. To better understand what we mean, consider this question: Is it more profitable to sell music exclusively in an album format, or will labels make more profit by selling both bundled albums and unbundled singles? That was the question facing the music industry in the late 2000s, as iTunes was becoming popular. The conventional wisdom in the industry was clear: Digital singles are bad for business, possibly even worse than digital piracy. "Stealing music is not killing music," said the MTV co-founder Robert Pittman. "When I talk to people in the music business, most of them will admit the problem is they're selling songs and not albums. I mean, you do the math."[20]

On the surface, the math seemed pretty compelling. From 2002 to 2008, according to the International Federation of the Phonographic Industry, worldwide sales of physical music fell from $24.7 billion to $13.9 billion.[21] And sales of digital albums and singles only recovered a fraction of this loss, reaching $4 billion in 2009. Overall, revenue from recorded music, digital and physical, were down 28 percent in that same period, from $24.7 billion to $17.9 billion. What was happening seemed obvious: Consumers who used to buy CD "albums" for $15–$20 were now buying only a few singles and spending only a few dollars, right? Maybe, but maybe not. What if unbundling singles from albums, instead of reducing revenue by enabling consumers to spend less for the music they wanted, *increased* revenue by attracting new consumers? Should the music industry stop selling singles and go back to selling only albums? We analyzed these questions in early 2009, partnering with a major music label to run an experiment.[22] Specifically, we began by working with the label to select of 2,000 of their best-selling singles. We then randomly increased the price of some of these singles from $0.99 to $1.29 on one of the label's major digital-sales platforms. That gave us an opportunity to see how sales of the newly re-priced singles would change, how sales of other singles taken from the same album would change, and how album sales would change. When we ran the numbers and did some econometric modeling, we got a clear answer: Artists and labels both make more money when they sell their content as digital singles instead of only selling digital albums.

Place

In the first quarter of 2015, according to Nielsen, American adults watched 16 fewer minutes of live television per day than they had two years earlier.[23] Some in the industry have attributed this drop to Nielsen's inability to keep up with new ways consumers are viewing content. Viacom's CEO, Philippe Dauman, said that Nielsen's measurement services have "not caught up to the marketplace."[24] Nielsen's CEO, Mitch Barnes, shot back, implying that the problem wasn't Nielsen but the poor quality of television content: "Sometimes they are using us as a scapegoat. When ratings are down, you'd hate it to be because your programming maybe fell a little bit behind. You'd much rather have it be because maybe somebody isn't doing everything that you think they ought to do."[25] There is a third possibility, however. Ratings may be down because consumers are spending more time on the Internet. (The time Americans spent on the Internet increased by 42 minutes per day from 2013 to 2015.)

Does increased Internet consumption reduce television viewing? That's the question our Carnegie Mellon colleague Pedro Ferreira set out to answer in 2015 by conducting an experiment with a major cable company that doubles as an Internet service provider. In the experiment, Ferreira and his co-authors randomly chose 30,000 of the company's subscribers and conducted an experiment that gave half of them free access to premium television channels with time-shift capabilities, and compared how their television and Internet use changed relative to the remaining (control-group) users, who didn't receive the free offer. They discovered a strong relationship between increased Internet use and decreased television viewership: Relative to the behavior of those in the control group, users who adopted the free television channels reduced their use of the Internet significantly.[26] The experiment suggests that the more people use the Internet, the less they watch TV—a sobering thought for the television industry.

Price

Data can also play a significant role in informing managerial decisions about price—decisions that previously were made loosely, primarily by "gut feel." Consider the following example: In 2000, we called a marketing executive at a major publishing house in the hopes of learning how much the publisher's sales changed when the price of a typical book changed—something that economists refer to as *price elasticity*. We started the call by asking what we thought was a simple question: "What's a typical price elasticity for a book?" Long silence on the other end of the line. Maybe we hadn't been specific enough. "A hardcover book?" Another long silence. Maybe he didn't understand the economic terminology. "If you were to reduce the price of a hardcover book by 10 percent, how much would you expect sales to increase?" A third long silence. Eventually the executive told us that the publishing industry did very little quantitative analysis before setting prices. Prices were set on the basis of industry norms, competitors' decisions, and a lot of intuition. For a long time, that was a perfectly reasonable way to set prices. The industry was relatively stable and could get away with relying on standard practices when setting prices. But are the prices that worked well for selling hardcover books, CDs, and DVDs ten years ago the right prices for selling e-books, digital albums, and movie downloads today? How should firms go about determining the right prices for their content in new digital marketplaces?

These are genuinely difficult questions because consumers now have a dizzying variety of movies and TV shows competing for their attention—and their dollars. Consumers can buy or rent DVDs. They can subscribe to many different cable offerings. They can buy or rent digital content from iTunes, Amazon, and Netflix, or they can stream it from all sorts of sources, both legal and illegal. Changing the price in any one of these channels has implications for sales in other channels. Without proper coordination among these channels, where decisions are now often made independently, pricing by one division might well hurt the others. Multi-channel pricing decisions, in other words, are hard.

"Gut feelings" are no longer good enough in this sort of constantly evolving environment. To develop an effective price strategy, entertainment firms will have to develop a thoughtful data-driven approach that will allow them to factor in the many variables now at play in the pricing equation.

For example, pricing has to be not only channel-dependent but also time-dependent, which adds significant complexity. As we've discussed, when selling information goods the best way to generate the most revenue over time is to use price discrimination—to charge a high price for high-value consumers (who typically want to get the content immediately after release) and a lower price for lower-value consumers (who are willing to wait). The trick here is developing a model that allows for opportunistic pricing—that lowers prices when demand is weak and raises them when demand is strong. The optimal price of a book, a song, or a movie will change over time, so the question becomes how to extract the most value in a rapidly changing market.

Let's consider a specific problem. Consumers have traditionally been willing to pay $15–$20 to buy a DVD. But what are they willing to pay for movies on digital platforms? The best way to answer this question is with a data-driven experiment—which is exactly what we did in partnership with a major studio. In the study, we took a large number of the studio's older "catalog" titles and experimentally lowered their prices on a major online distributor's platform. During our experiment, some prices dropped from $9.99 to $7.99, some to $5.99, and some to $4.99, leaving the rest unchanged, as a control. We found that online consumers were extremely sensitive to price. Often, if we lowered prices by half, consumer demand tripled or quadrupled. Some of this increased demand came at the expense of rentals or sales on other digital platforms, but even so it was clear that lower online prices led to higher total sales, higher revenue, and higher profit for the studio.[27]

The other entertainment industries can very profitably optimize prices too. The experiment that we ran to determine whether a label's digital singles were helping or hurting album sales also allowed us to determine optimal pricing strategies for the label's singles and albums. In general, what we found was that prices for singles were about 30 percent too low, and prices for albums were about 30 percent too high. After implementing these price changes, the executive we worked with told us that the label realized "tens of millions of pounds of profit a year" from making this change. "The whole industry was wrong," the executive said. "We thought that we knew the right price, and we were just wrong."

The point is that using experimentation and data analytics to optimize prices can help firms increase their profits, sometimes by quite a lot. This is good news for the entertainment industries. But because optimal prices will fluctuate over time, across a wide range of products, entertainment firms will need to continually optimize their prices, which brings us to the bad news. Right now, the new platform companies are in a much better position to optimize prices than the majors. Amazon, for example, sells 100 million unique items, and it sets prices for these products entirely with software—software that runs experiments, tests consumer responses, and updates prices when necessary. And because this all happens automatically, Amazon's prices might occasionally be a few percentage points off, but they are never off by 30–50 percent.

The platform companies can also optimize prices by using their market power to enforce pricing policies that benefit their business. For example, our pricing experiment showed that the label could increase its profit even more if it could choose from many different prices for singles depending on their popularity or genre. The problem was that the platform company preferred the simplicity of $0.99 or $1.29 pricing as a way to increase platform adoption and to sell more hardware, and the labels weren't powerful enough to force it to change.

Promotion

Data analysis can also help entertainment firms improve the effectiveness of their marketing and advertising campaigns. Those campaigns often consume a significant portion of the money that firms spend on their content (sometimes up to 40 percent of the total cost of a movie, for example), but most are still untargeted. They simply involve placing advertisements in as many channels as possible to inform as many consumers as possible about the content—a strategy one executive we spoke to referred to as "spray and pray."[28] Before criticizing this approach, it's important to realize that studios use the "spray and pray" approach, not because they're unsophisticated but because for a long time that's all they could do. In traditional advertising channels, it's extremely difficult to measure consumer response, and without a good way to measure who was responding to which advertisement, the studios had no way of measuring how effective their advertising campaigns were. Nor did they have any way of determining the counterfactual—that is, what sales would have been had they not run their ads.

With the growth of the Internet, however, new opportunities are emerging that can help firms to develop better-targeted—and more profitable—campaigns. Consider the experiment we ran in partnership with a major studio and Google's ad-sales team to study the value of online advertising—specifically, the value of targeting consumers who had previously watched movie trailers for some of the studio's catalog movies. We divided the United States into more than 400 regions and then randomly assigned what advertisements were shown to users in each region. In one third of the regions, when users viewed a clip related to the promoted movie, we displayed an advertisement encouraging them to buy the movie at an online digital storefront. In another third of the regions, we displayed advertisements on the same set of clips, but this time we only showed advertisements to users who had also watched the trailer for the movie on a previous visit. The remaining third of the regions served as a control group and was shown no advertisements.

We got some very striking results. Even after taking into account the higher cost Google charges to target users who had previously viewed a trailer for the advertised movies, we found that targeted ads of this sort generated profits for studios that were four to five times the profits generated by untargeted ads. Knowing what content individual consumers have explored in the past turns out to be a strong predictor of what they're likely to buy in the future.

The challenge for the studios and the other majors is that they don't control this important customer information. The ability to target users who had watched the trailer for the movie was available to us because we had gotten Google's permission to run our experiment. Think of all the information that Google, Amazon, and other data-driven companies possess, and what an advantage it gives them when it comes to targeting consumers.

Online placement represents another promising promotional opportunity—one that has the potential to become much more important than brick-and-mortar placement. Sites such as iTunes and Amazon can powerfully influence movie-buying decisions, for example, by placing certain films on their websites, by directing traffic to those films, or by sending out targeted e-mail messages to customers who they know are likely to be interested. And the nature of digital "inventory" means the firms can respond to brief spikes in demand much more dynamically than physical stores can. Consider, for example, that on a Friday the 13th online sales of the horror classic *Friday the 13th* spike, but sales in physical stores don't. Why? Because physical retailers have to place orders for titles well in advance, and can't respond nimbly to trending developments. Not only that; if physical retailers place advance orders for such titles, they then have to use up precious retail space to stock them for a short period, and after which they have to retrieve the unsold physical stock and send it back to the studio. This is why you don't see physical retailers changing their inventories for the two or three days before a Friday the 13th. Their distribution mechanism and space limitations mean they can't react quickly enough to take advantage of the temporary money-making opportunity.

On the Internet, however, you can respond almost instantaneously to these brief changes in demand. You can look at your data, and you can even use machine-learning techniques, to quickly discover (and in some cases predict) the changes in demand—and then you can immediately move high-demand titles to the "front" of your virtual store. None of that is possible with "gut feel." The only way you can make it happen is if you have the permission of the firm that owns the distribution channel. This puts Amazon, for example, at a huge advantage, because it can name its price for promoting content on its site. Is $100,000 for temporary placement on Amazon's front page a great deal or a swindle? Without running experiments and studying data on customer responses you can't answer that question. But Amazon can, giving it a huge informational advantage when bargaining for advertising placement.

• • •

We began this chapter with the story of Harrah's because it showed that companies can change their organizational structure to leverage new data-based analytics and management techniques. That's an important message for the majors in the entertainment industries to hear, but organizational change is only part of the battle they will have to fight in the years ahead. Most entertainment firms don't know their customers as individuals. Harrah's was able to use its casinos to interact with customers directly, but historically the entertainment majors' most important interactions with customers have occurred through such third-party intermediaries as movie theaters, music shops, and bookstores—brick-and-mortar institutions that posed little threat to the majors' business. But increasingly, as we've been arguing, these interactions are occurring through large data-driven aggregators, such as Amazon, iTunes, Netflix, and Google, that directly observe the behavior of each of their customers and are already organized around data analytics and evidence-based management. They're much better positioned to collect detailed data about customers' behavior, to estimate a customer's lifetime value, to run experiments to evaluate the effectiveness of various marketing strategies, and to direct specific promotions to individual customers in order to increase their loyalty. In addition, they hold an increasingly powerful position within the entertainment supply chain, and in some cases they're using their platforms and their data on customers to create content of their own.

That's the bad news. But there is good news for the entertainment industries, which we'll be discussing in detail in our next chapter. In brief, the majors already have at their disposal many of the tools they need to interact directly with their customers and to develop new, evidence-based management strategies. If they start using them, they should be able to respond effectively to the threat posed by their new competitors.

11   The Show Must Go On

Get closer than ever to your customers. So close that you tell them what they need well before they realize it themselves.

Carmine Gallo, *The Innovation Secrets of Steve Jobs: Insanely Different Principles for Breakthrough Success* (McGraw-Hill, 2010)

If studios want to thrive in the era of Google, Amazon, and Netflix, they are going to have to think differently about communicating with their customers. In order to do that, they are going to have to make gathering and analyzing data on customers a priority.

Consider the instructive story of how Steve Jobs revived Apple and transformed it into one of the most successful businesses in the world. The story in general is well known, but we'd like to focus on one aspect of it that gets less attention than it should: how Apple used connections with customers and data on customers to turn itself around.

In 1997, when Jobs returned to Apple, the company was struggling. It had barely 4 percent of the computer market, its stock price had just hit a twelve-year low, and many industry experts were predicting it would soon have to fold. On October 6, Michael Dell said at a Gartner Symposium that if he were in charge of Apple he would "shut it down and give the money back to the shareholders."[1]

One of Apple's big problems was that, with a small share of the market, it couldn't reach its customers directly. It relied on third-party retailers, including Sears, Best Buy, Circuit City, and OfficeMax, to sell its computers to customers, and those retailers had no incentive to create customer loyalty to Apple's brand. Their salespeople knew little about Apple products and, in fact, would often steer customers away from Apple products and toward cheaper Windows computers. Apple's products were relegated to poorly trafficked and poorly maintained parts of the stores. Many people simply weren't aware of the benefits that Apple computers had to offer. Jobs recognized this as a major obstacle. His plan for turning Apple around involved delighting customers, but it isn't easy to delight customers if you have no way of reaching them or even knowing who they might be.

Jobs came up with a crazy solution: Apple would build its own stores. Why was that crazy? Well, for one thing, retail-store space was expensive. It seemed almost ludicrous to propose that the way to compete with Dell in a slim-margin business was to make a huge investment in retail stores. That approach certainly hadn't worked for Gateway. In January of 2001, four months before Apple would open its first retail store, Gateway, suffering in competition with Dell, had been forced to close 27 of its retail stores.[2]

Not surprisingly, the business press openly mocked Apple's plan. In an article titled "Sorry Steve, Here's Why Apple Stores Won't Work," *Business Week* suggested "Maybe it's time Steve Jobs stopped thinking quite so differently."[3] "Apple's problem," the company's former chief financial officer Joseph Graziano said, "is it still believes the way to grow is serving caviar in a world that seems pretty content with cheese and crackers." David Goldstein, a retail consultant for Channel Marketing Corporation, summed up the conventional wisdom: "I give them two years before they're turning out the lights on a very painful and expensive mistake."[4]

Today, Apple has 453 retail stores in 16 countries. In its earnings announcement for the first quarter of 2015, Apple reported that 500 million consumers visited its retail and online stores, and that the retail stores generated nearly $4,800 in annualized sales per square foot—more than those of any other retailer in the United States.[5] According to *Forbes*, Apple has 50,000 retail employees worldwide, who serve, on average, a million customers per day.[6] Apple's retail operation is now worth more than all of Apple was worth in 2001.

Apple's retail stores succeeded largely because, instead of just pushing products as most computer retailers did, Apple focused on the consumer experience. In particular, it designed its stores not around product lines but around the needs of its consumers, showing them how they could use Apple products to listen to music, take pictures and videos, and watch movies. Significantly, Apple also staffed its stores with friendly people who could guide customers through their purchases and teach them how to use what they were buying. That part of the story is well known. What gets less attention is that Apple used its retail stores to take control of how its products were displayed and marketed to its customers—and it did that by using data.

Experiments and data influenced every aspect of Apple's stores. The company spent lavishly on mock-ups of various store layouts and used feedback from customers to refine their design. It interviewed customers about their best customer-service experiences and used what it learned to design the Genius Bar. It studied market data and demographic data in order to locate its stores conveniently so that new users to the Mac platform (most of them Microsoft users) had to gamble not with "20 minutes of their time," as Steve Jobs put it, but only "20 footsteps of their time."[7]

Apple used data to create its stores and the experiences people had in them—but then it used the stores to create a data feedback loop from the customers back into the company. The stores—from the Genius Bar and the one-to-one training areas to the sophisticated technology used to identify the physical location of in-store customers[8]—were carefully designed to collect data about customers that Apple could then feed back into how it designed and marketed products and served customers. How were customers using Apple devices? What did they like best about them? What about the devices did customers find frustrating? What devices were breaking, in what way, and after how long? What did people want? What did they need? The retail stores put Apple directly in touch with customers, and the company used the stores expertly to collect information that helped them understand and serve their customers' needs.

• • •

We aren't suggesting that Paramount Pictures or Universal Music open stores in swanky shopping districts all over the world. However, in a world in which direct connections with customers are becoming increasingly important, we do believe that the major entertainment companies will put themselves at a strategic disadvantage if they rely exclusively on third-party distributors to showcase their products to customers. To compete, they will have to make the kind of transformation that Apple did—a shift that involves communicating value through direct connections with customers rather than through intermediaries, and that involves collecting and using data on customers to understand and serve customers' individual needs. How might this work in practice?

Let's consider the motion-picture industry. Some of the data the studios will need to connect with their customers is readily available on various social media sites and can be accessed relatively easily. Legendary Pictures, for example, has invested in an analytics division that aggressively collects consumer data from every source they can find it in—Twitter, Facebook, Google, ticketing data—and then uses it to better reach the right customers with the right promotional message. Without this kind of targeting, Legendary's CEO Thomas Tull has said, studios are wasting money by spending as much money on marketing *The Dark Knight* to eighty-year-old women as they spend on marketing it to teenaged boys.[9]

However, as we have discussed previously, the most valuable data for the studios are proprietary and closely held by Apple, Amazon, Google, and Netflix. A simple approach for the studios to adopt would be to prioritize getting access to data on individual customers when they negotiate with their distributors. In fact, many studios have already begun to press their distributors for more detailed data. But the studios will face obstacles as they try to adopt this approach. They will have to make significant concessions to powerful online distributors if they want to be able to gain access to their customers. Even if the studios manage to negotiate that access, they will still only be able to see customers' behavior for their own content, whereas online distributors will be able to observe customers' behavior across all content on their platforms. The studios also will not be able to benefit from the sort of direct testing and experimentation that we discussed in the previous chapter. And, maybe most important, if the studios rely on these distributors to maintain access to customers, they will increasingly be relying on their competitors for access to strategic information.

It's important to realize that Amazon, Netflix, and Google are vertically integrating upstream into the original content business in large part so they will be less reliant on the major studios for access to the studios' content. For that reason alone, the studios should turn the tables on their new rivals: vertically integrating downstream into direct distribution so they will be less dependent on the distributors for access to customers. The most straightforward way to do that, at least for content with a strong brand, would be to invest in attracting consumers to the studios' existing online portals. Here the studios could adopt J. K. Rowling's strategy with Pottermore and create communities in which creators can share additional information with fans. This would enable the studios to keep track of consumers' behavior across the various communities in their roster, and to use that information to promote content directly to fans. The main obstacle facing that approach is that, as we have discussed, consumers have a strong desire for simplicity. They may be hesitant to learn how to use multiple websites and unwilling to maintain multiple logins across multiple content sites. Likewise, if each studio maintains its own separate platform, it will be able to observe only how consumers interact with the studio's own content.

A more ambitious approach, and one that we believe holds the most promise for success, involves the studios' forming a strategic partnership to invest in a common platform that would allow each firm direct, targeted access to its customers while enabling each firm to see and understand customers' behavior across multiple firms' content. And such a platform already exists. In March of 2007, three major studios—21st Century Fox, NBC Universal, and Walt Disney Studios/ABC Television—announced that they had teamed up to create "the largest Internet video distribution network ever assembled."[10] Today that network, called Hulu, is the fourth-most-popular video-streaming platform in the United States, just behind Amazon's Instant Video service.[11]

Unfortunately, Hulu's success has created an almost intractable problem. The more popular Hulu became, the more likely it was to reduce the profitability of the studios' established channels. Disputes arose almost immediately after its launch about which shows would be made available to Hulu, how many minutes of commercials Hulu should include in their streams, how many episodes of content would be available on the site, and how long Hulu would have to wait after the TV broadcast before making content available. These questions were all decided in the context of the current television business model, in which about half of a network's revenue comes from advertising dollars and the other half from "retransmission fees"—payments from cable companies for the rights to broadcast the networks' content. As a result, anything that might reduce a show's Nielsen ratings or retransmission fees was seen as a threat.

One way to deal with these conflicts would be to allow Hulu to operate completely independently, giving it the freedom to pursue new business models that might ultimately cannibalize the old ones. That was the idea behind a 2010 proposal for an initial public offering for Hulu, but the networks quickly killed it. "I don't know if any amount of money would have been enough to get them to give up control," the Wells Fargo analyst Marci Ryvicker observed in a *Fortune* interview.[12]

In February of 2011, in an effort to convince the networks to stop impeding Hulu's success, Hulu's CEO, Jason Kilar, posted a 2,000-word entry on Hulu's blog in which he lectured his bosses about the future of their business.[13] Kilar argued that there were too many advertisements on television, that consumers should be allowed to watch content on their own schedule, and that the business of forcing customers to buy huge cable packages of channels unrelated to their interests was dying, as was the business of charging cable companies huge retransmission fees to fill those channels with content. Kilar closed with a warning to network executives: "History has shown that incumbents tend to fight trends that challenge established ways and, in the process, lose focus on what matters most: customers."

The networks were not immediately grateful for Kilar's feedback. "Eighty to 90 percent of what he says is right," one unnamed network executive told the *Wall Street Journal*. "But why print that? Does he think we're going to say, 'Oh, thank you! You're right! We'd never thought of that! Let's give away retrans [retransmission fees]!'?"[14] Another executive, quoted in the *Financial Times*, minimized Kilar's accomplishments and his ability to work within the existing system: "If I were given billions of dollars worth of programming, I too could probably build a business. But I know that in order to build a long-term, viable business I would have to do so in a way that works for everybody."[15] A third executive was even more blunt: "These are clearly the musings of an elitist who is obviously disconnected from how the majority of America watches television."[16]

In the end, many industry observers feel that a vibrant new streaming platform isn't compatible with the studios' existing business models. "They don't want it to succeed," the media analyst James McQuivey wrote. "It's media economics—if it succeeds, it will do so by cannibalizing the currency of the television media business today, which is television ratings. So they made a decision to make sure Hulu didn't get too good or become too successful."[17]

• • •

The gut instinct to disadvantage digital channels to avoid cannibalizing existing revenue streams may seem perfectly reasonable under the old rules of the entertainment business. After all, people who can't consume digitally will have no other choice but to buy the physical product, right?

That's a question we examined recently using digital and DVD sales data collected from 2012 and 2013. Before 2012, the conventional wisdom in most of the motion picture industry was that delaying the release of movies on iTunes and other digital channels would protect the studios' valuable DVD revenue. However, from 2012 to 2013 several studios started to experiment with releasing their titles digitally at the same time as, or in some cases before, the DVD release. This shift in strategy at the studio level allowed us to analyze how sales in both channels changed when consumers could choose between DVDs and iTunes downloads.[18] The data showed that delaying digital availability has a huge downside and almost no upside. When digital movies were released after the DVD release date, digital sales were cut by almost half and there was no statistical increase in DVD sales.

This aligns well with what we found when we studied what happened after digital content was made available in other contexts: on the Kindle e-book store (chapter 3), through Hulu television streaming (chapter 6), or as iTunes television downloads (chapter 8). Although digital channels are undoubtedly reducing physical consumption at a macro level, individual firms can do very little to forestall this trend for individual titles by delaying the digital availability of their content. Delaying digital availability has almost no effect on physical sales, because digital consumers are already gone. They are either pirating the content or consuming other types of content on Netflix, iTunes, Amazon, or YouTube.

Delaying digital availability is also risky for firms because without strong digital platforms the majors will not be able to tap into the many important benefits of digital distribution. We have identified five benefits of digital distribution that we feel are particularly significant. We have already discussed two of them extensively: the ability to better evaluate the potential market for content and the ability to more efficiently promote this content to consumers. Netflix has exploited both of these advantages, along with a third: the ability to conduct and learn from detailed experiments about how consumers respond to content. In her book *Netflixed: The Epic Battle for America's Eyeballs*, Gina Keating describes how the company has used its website to understand its customers' needs:

[Netflix] designed the Web site to double as a market research platform that could display multiple versions of a page or feature to test groups of customers and gather detailed data on their reactions and preferences. A typical A-B test involved measuring the effect of a red logo (choice A) versus a blue logo (choice B) on acquiring a customer, and their lifetime value, retention rate, and usage. ... The constant testing, gathering of consumer input, and subsequent adjustments to the site formed an ongoing conversation between Netflix and its customers that would provide a crucial advantage in the coming battle with store-based renters.

Of course, competitors can simply copy those design decisions, and at the time Blockbuster.com did just that, quickly incorporating elements of Netflix's design. But although Blockbuster.com copied the look of the Netflix site, it obviously couldn't copy the algorithms underpinning it. Without ongoing optimization of costs, of the matching algorithm, and of the market-research platforms, Blockbuster didn't have anything close to the whole picture.

Platform companies can do a very good job of customizing their marketing by using direct experiments, as we noted in the previous chapter. But they can also generate insights that wouldn't be possible without data on individual customers—our fourth benefit. For a long time, demographic data of the sort sold by Nielsen and other market-research firms were the only data on customers that were available, because demographic data were all that marketers could measure efficiently.[19] But demographic data tell you almost nothing about who people are or what they want to consume. In a world of digital interactions with customers, and of tremendous computing power, demographic data have become almost worthless for the purpose of making marketing decisions.

How valuable is it to know a customer's purchase history rather than only his or her demographic characteristics? Peter Rossi, Robert McCulloch, and Greg Allenby asked that question in a paper published in 1996.[20] At the time of their study—not long after grocery stores had adopted bar-code scanners and customer loyalty cards—marketers were just beginning to appreciate the importance of data on customers. Loyalty cards enabled grocery stores to see their customers as individuals for the first time. Rossi, McCulloch, and Allenby used data from a store's scanner data to compare the effectiveness of untargeted "blanket" coupon drops, coupons targeted on the basis of demographic data, and coupons targeted on the basis of a customer's purchase behavior. They found that knowing a person's demographic data increased the profitability of coupons by 12 percent relative to "blanket" couponing, and that adding an individual's purchase history increased the profitability of coupons by 155 percent relative to the blanket strategy.

A tenfold increase in marketing effectiveness is incredibly powerful at the scale at which Amazon works. But Amazon and other platform companies didn't stop there. Increasingly, they are basing their marketing decisions on their ability to observe a customer's behavior in real time. Amazon wants to customize its marketing on the basis of what you're currently searching for, what you're currently looking at, how often you click, and so on, because data on such behaviors help to answer the most salient marketing question of all: Why are you here right now?

The fifth benefit from observing detailed data on customers has to do with products rather than with customers. To clarify what we mean, let's go back to the grocery industry in the mid 1990s. Before grocery stores adopted customer loyalty cards, the conventional wisdom in the industry, based on a study conducted by the Food Marketing Institute, was that to improve their inventory management grocery stores should reduce the number of niche products they carried.[21] The H-E-B grocery chain, however, discovered that the Food Marketing Institute's study had ignored something fundamental to a store's profitability: that the most profitable customers were the ones most likely to purchase "niche" products. H-E-B recognized that if it eliminated these slow-selling niche products it might lose its most profitable customers, so its managers decided to stock *more* niche products.[22]

Might something similar be true in online purchasing—that the most profitable customers might be those who are interested in the least popular products? Seeking an answer to that question, we partnered with a major movie studio to analyze its customers' online purchases. The answer turned out to be Yes. On the whole, we found, sales skewed heavily toward blockbuster movies. That was no surprise. But it was a surprise that sales from the most profitable customers skewed heavily toward purchases of *obscure* movies. The most profitable customers were between 50 and 200 percent more likely than other customers to buy movies from the long tail.

• • •

The overarching lesson for the entertainment industries is that, to succeed in the future, companies are going to have to control the interface with their customers (and the resulting data about their customers' needs) in addition to controlling the production of content. That's what we have been arguing throughout this book.

As we discussed in chapters 1–4, for 100 years the major labels, publishers, and studios created value in their industries by using their size to manage two forms of scarcity: scarcity in the capacity of distribution and promotion channels and scarcity in the financial and technical resources necessary to create content. The majors were then able to capture value through business models that relied on maintaining control over how consumers were able to obtain content.

In chapters 5–9, however, we discussed how advances in computers, in storage equipment, and in the availability of worldwide digital-communication networks have made these scarce resources more and more plentiful. Low-cost production equipment has enabled almost anyone to become a content creator, and digital channels have provided a wealth of new opportunities to promote and distribute content—in turn, creating enormous amounts of value for consumers. Technological change has also affected the processes used to capture value in entertainment markets. Digital piracy has made it increasingly difficult for entertainment firms to maintain an artificial scarcity in consumers' access to content—once something gets released into the digital ecosystem, controlling its spread is almost impossible. But digitization has also made possible new ways of capturing value online through the convenience, personalization, and immediate gratification of on-demand content.

In chapter 10 and in this chapter, we argued that the keys to using these new tools to create and capture value in the entertainment industries will come from two new scarce resources: knowledge of customers' needs and the ability to manage customers' attention. As we discussed in the chapter 10, entertainment firms will have to prioritize data-driven decision making if they want to understand the needs of individual customers. That will require major investments in organizational change and a willingness to develop new forms of organizational talent. But if entertainment firms want to manage customers' attention, they will also have to make bold bets on new distribution platforms that will allow them to interact with their customers directly.

Of course this transition will be difficult for the entertainment industries. But we're optimistic about their future. That's because the steps we have called for above are the same steps that have always defined success in the entertainment industries: a willingness to take big risks on emerging opportunities, a desire to invest in new talent, a passion for finding creative ways of connecting artists with audiences, and the skill necessary to take a grand concept and make it a reality. One way or another, the show must—and will—go on.

Notes

Chapter 1

1. Source: http://bigstory.ap.org/article/netflix-shuffles-tv-deck-house-cards

2. Source: http://www.vulture.com/2014/05/kevin-reilly-on-fox-pilot-season.html

3. Nellie Andreeva, "Focus: 2009–2010 Pilot Season—Back on Auto Pilot," *Hollywood Reporter*, March 6, 2009, as quoted by Jeffrey Ulin in *The Business of Media Distribution* (Focal, 2010).

4. Source: Ted Sarandos, speech to the 2013 Film Independent Forum (http://www.youtube.com/watch?v=Nz-7oWfw7fY)

5. Ibid.

6. Ibid.

7. Source: http://www.nytimes.com/2013/01/20/arts/television/house-of-cards-arrives-as-a-netflix-series.html

8. Source: http://www.aoltv.com/2011/03/18/netflix-builds-house-of-cards-kevin-spacey/

9. That number represents 2 percent of Netflix's entire customer base. Source: http://tvline.com/2014/02/21/ratings-house-of-cards-season-2-binge-watching/

10. Of course, Netflix viewers weren't the only ones avoiding commercials. According to TiVo, 66 percent of TiVo's viewers of *The Walking Dead* and 73 percent of its viewers of *Mad Men* used the DVR to skip commercials—much to the consternation of advertisers who had paid $70,000–$100,000 to place 30-second commercials on those series.

11. Source: http://www.nytimes.com/2013/01/20/arts/television/house-of-cards-arrives-as-a-netflix-series.html

12. Source: http://www.hollywoodreporter.com/video/full-uncensored-tv-executives-roundtable-648995

13. Source: https://www.youtube.com/watch?v=uK2xX5VpzZ0

14. The trailers consisted of brief promotional advertisements for the show.

15. Source: http://www.nytimes.com/2013/02/25/business/media/for-house-of-cards-using-big-data-to-guarantee-its-popularity.html

16. Source: http://variety.com/2014/digital/news/netflix-streaming-eats-up-35-of-downstream-internet-bandwidth-usage-study-1201360914/

17. Source: http://stephenking.com/promo/utd_on_tv/

18. Source: http://www.nytimes.com/2012/08/05/sunday-review/internet-pirates-will-always-win.html

19. We will have more to say about the economics of bundling in chapter 3.

20. Source: http://www.gq.com/story/netflix-founder-reed-hastings-house-of-cards-arrested-development

21. Source: https://www.sandvine.com/downloads/general/global-internet-phenomena/2011/1h-2011-global-internet-phenomena-report.pdf

22. Source: http://variety.com/2015/digital/news/netflix-bandwidth-usage-internet-traffic-1201507187/

Chapter 2

1. The historical discussion that follows derives primarily, and sometimes closely, from three sources: Jan W. Rivkin and Gerrit Meier, BMG Entertainment, Case 701-003, Harvard Business School, 2000; Pekka Gronow and Ilpo Saunio, *An International History of the Recording Industry* (Cassell, 1998); and Geoffrey P. Hull, *The Recording Industry* (Routledge, 2004).

2. Source: http://historymatters.gmu.edu/d/5761/

3. Rivkin and Meier, BMG Entertainment, p. 3.

4. Ibid., p. 4.

5. Gertrude Samuels, "Why They Rock 'n' Roll—And Should They?" *New York Times*, January 12, 1958.

6. "Yeh-Heh-Heh-Hes, Baby," *Time* 67, no. 25 (1956).

7. Samuels, "Why They Rock 'n' Roll."

8. Ibid.

9. R. Serge Denisoff and William D. Romanowski, *Risky Business: Rock in Film* (Transaction, 1991), p. 30.

10. "Rock-and-Roll Called 'Communicable Disease,'" *New York Times*, March 28, 1956.

11. See, for example, Reiland Rabaka, *The Hip Hop Movement: From R&B and the Civil Rights Movement to Rap and the Hip Hop Generation* (Lexington Books, 2013), p. 105; Glenn C. Altschuler, *All Shook Up: How Rock 'n' roll Changed America* (Oxford University Press, 2003), p. 40; Peter Blecha, *Taboo Tunes: A History of Banned Bands and Censored Songs* (Backbeat Books, 2004), p. 26; Linda Martin and Kerry Segrave, *Anti-Rock: The Opposition to Rock 'n' Roll* (Da Capo, 1993), p. 49.

12. "Boston, New Haven Ban 'Rock' Shows," *New York Times*, May 6, 1958.

13. Samuels, "Why They Rock 'n' Roll."

14. Gronow and Saunio, *An International History of the Recording Industry*, pp. 193–194.

15. William Goldman, *Adventures in the Screen Trade* (Warner Books, 1983), p. 39.

16. BMG Entertainment, p. 8.

17. International Federation of the Phonographic Industry, Investing in Music: How Music Companies Discover, Nurture and Promote Talent, 2014, pp. 7–9.

18. Robert Burnett, *The Global Jukebox*, as cited in BMG Entertainment.

19. Steve Knopper, *Appetite for Self-Destruction: The Spectacular Crash of the Record Industry in the Digital Age* (Free Press, 2009), p. 202.

20. Michael Fink, *Inside the Music Industry: Creativity, Process, and Business* (Schirmer, 1996), p. 71.

21. Hull, *The Recording Industry*, p. 186; quoted in "Payola 2003," *Online Reporter*, March 15, 2003.

22. Source: Erik Brynjolfsson, Yu Hu, and Michael Smith, "Consumer Surplus in the Digital Economy: Estimating the Value of Increased Product Variety," *Management Science* 49, no. 11 (2003): 1580–1596.

23. Source: our calculations, based on http://www.boxofficemojo.com/studio/?view =company&view2=yearly&yr=2000

24. See, for example, Albert N. Greco, Clara E. Rodriguez, and Robert M. Wharton, *The Culture and Commerce of Publishing in the 21st Century* (Stanford University Press, 2007), p. 14.

Chapter 3

1. Source: http://online.wsj.com/news/articles/SB125427129354251281

2. Source: http://shelf-life.ew.com/2009/10/23/stephen-king-ebook-delay-price-wa/

3. Jeffrey A Trachtenberg, "Two Major Publishers to Hold Back E-Books," *Wall Street Journal*, December 9, 2009.

4. The other assumption these publishers seem to be making is that the high-priced hardcover books have higher margins than the lower-priced e-books. In practice, however, given the costs of printing and distributing hardcover books, we found that the margins for these to products are actually very close to each other.

5. Although the details are beyond the scope of this book, in the research, we attempted to test for whether the timing of the event was truly exogenous to our experiment and whether the schedule of book releases was uncorrelated with expected sales. We refer interested readers to a working paper by Hailiang Chen, Yu Jeffrey Hu, and Michael D. Smith titled The Impact of eBook Distribution on Print Sales: Analysis of a Natural Experiment, which is available from http://ssrn.com/abstract=1966115.

6. As we will discuss in more detail below, these characteristics are shared by the content of motion pictures and that of music. For example, the costs of making and promoting a movie can exceed $100 million, but the marginal cost of manufacturing a movie on a DVD is about $4.10 (see "The Hollywood Economist: The Hidden Financial Reality Behind the Movies," Epstein. 2012. Melville House Publishing, Brooklyn, NY.) and is essentially zero for additional copies of digital movies sold by iTunes.

7. More precisely, just a bit less than their willingness to pay.

8. Arthur C. Pigou, *The Economics of Welfare*, fourth edition (Macmillan, 1932).

9. Another problem with first-degree price discrimination strategies is that most customers consider them unfair—they don't see why they should be forced to pay more for a product just because they might be willing to.

10. To isolate the effect of the pay-cable broadcast from any effect of removing the movies from other channels, we used the fact that, according to their contracts, studios were required to remove their content from other "competing" channels on the first day of the month in which a movie was shown on the pay-cable network, but that the actual broadcast date typically occurred on the first, second, third, or fourth weekend of the month. For example, in our paper (Anuj Kumar, Michael D. Smith, and Rahul Telang, "Information Discovery and the Long Tail of Motion Picture Content," *Management Information Systems Quarterly* 38, no. 4 (2014): 1057–1078) we observe that "in March 2011, the movies *Robin Hood*, *MacGruber*, *Cop Out*, and *Just Wright* premiered on HBO. These movies were all removed from iTunes and cable

pay-per-view channels on March 1 and were first broadcast on HBO on March 5, 12, 19, and 26 respectively." This difference between the date of removal and the date of broadcast allowed us to separately identify the effect of the removal on iTunes and pay-per-view from the effect of the broadcast on HBO.

11. The details of why we believe this change is causally related to the HBO broadcast, and not merely correlated with movie characteristics or changes in promotion and distribution, are beyond the scope of this book. We refer interested readers to Anuj Kumar, Michael D. Smith, and Rahul Telang, "Information Discovery and the Long Tail of Motion Picture Content," *Management Information Systems Quarterly* 38, no. 4 (2014): 1057–1078.

12. We discuss other strategic implications of bundling content in more detail in chapter 8.

13. In chapter 5 we expand on this concern and provide empirical evidence that it may already have harmed investment in some markets.

Chapter 4

1. The managerial concepts we will discuss in this chapter are aligned with a variety of managerial theories, including Joseph Schumpeter's theory of creative destruction, Clay Christensen's theory of disruptive innovation, and Richard Foster's concept of attacker's advantage. We apply these concepts in a setting where multiple simultaneous changes make it harder for incumbents to evaluate the magnitude of the problem, and where economies of scale enjoyed by entrants exacerbate the risk of delayed action by incumbents.

2. Source: http://www.prnewswire.com/news-releases/att-launches-a2b-music-with-the-verve-pipe--a-trial-for-the-delivery-of-music-over-the-internet-77352797.html

3. AAC compression was used to encode a2b files. Patents on the technology were held by AT&T Bell Laboratories, Fraunhofer IIS, Dolby Laboratories, and Sony Corporation. See Karlheinz Brandenburg, "MP3 and AAC Explained," presented at AES 17th International Conference on High Quality Audio Encoding, 1999 (available at http://www.aes.org/e-lib/browse.cfm?elib=8079).

4. Gronow and Saunio, *An International History of the Recording Industry*, p. 211.

5. Our synopsis of the Britannica case closely follows the account laid out by Shane Greenstein and Michelle Devereux in The Crisis at Encyclopaedia Britannica, Case Study KEL251, Kellogg School of Management, 2006 (revised 2009).

6. Ibid., p. 2, citing Randall E. Stross, *The Microsoft Way*.

7. Ibid., p. 5, note 21, quoting Philip Evans and Thomas S. Wurster, *The Microsoft Way*.

8. Ibid., p. 17, citing Robert McHenry, "The Building of Britannica Online" (http://www.howtoknow.com/BOL1.html).

9. Ibid., p. 17, citing Robert McHenry, "The Building of Britannica Online" (http://www.howtoknow.com/BOL1.html).

10. Ibid., p. 17, citing Stross, *The Microsoft Way*.

11. Ibid., p. 7, citing Dorothy Auchter, "The Evolution of *Encyclopaedia Britannica*," *Reference Services Review* 27, no. 3 (1999): 291–297.

12. Ibid.

13. Matt Marx, Joshua S. Gans, and David H. Hsu, "Dynamic Commercialization Strategies for Disruptive Technologies: Evidence from the Speech Recognition Industry," *Management Science* 60, no. 12 (2014): 3103–3123.

Chapter 5

1. Here we are using the standard definition of "long tail," which, according to the Oxford Dictionaries website, refers to "the large number of products that sell in small quantities, as contrasted with the small number of best-selling products." (See http://www.oxforddictionaries.com/us/definition/american_english/long-tail.)

2. Erik Brynjolfsson and Michael Smith, "Frictionless Commerce? A Comparison of Internet and Conventional Retailers," *Management Science* 46, no. 4 (2000): 563–585.

3. This point was first made raised by the noted economist John Kenneth Galbraith, who reviewed *The Winner-Take-All Society* in 1995 for the *Harvard Business Review*. In the review, titled "The Winner Takes All … Sometimes," Galbraith wrote: "Athletics, which is the authors' starting point and to which, rather significantly, they frequently return, is programmed to produce a clear-cut winner; not so for many other activities, even where there is considerable market concentration."

4. For more on our methods and results, see Erik Brynjolfsson, Yu Hu, and Michael Smith, "Consumer Surplus in the Digital Economy: Estimating the Value of Increased Product Variety," *Management Science* 49, no.11 (2003): 1580–1596.

5. Source: Bowker, cited in *Statistical Abstract of the United States: 2004–2005* (Government Printing Office, 2004), p. 721, table 1129.

6. The increase in books in print is, in many ways, interesting in and of itself. According to Bowker (http://www.bowkerinfo.com/pubtrack/AnnualBookProduction2010/ISBN_Output_2002-2010.pdf), the number of new titles printed per year increased from 562,000 in 2008 to 3.1 million in 2010. Much of this growth was driven by "non-traditional" (typically self-published) titles. The percentage of non-traditional titles increased from 13 percent in 2002 to 92 percent by 2010.

7. See Luis Aguiar and Joel Waldfogel, Quality, Predictability and the Welfare Benefits from New Products: Evidence from the Digitization of Recorded Music, working paper, University of Minnesota, 2014.

8. Anita Elberse, "Should You Invest in the Long Tail?" *Harvard Business Review* 86, no. 7/8 (2008): 88–96.

9. Glenn Ellison and Sara Fisher Ellison, Match Quality, Search, and the Internet Market for Used Books. working paper, Massachusetts Institute of Technology, 2014.

10. McPhee's original 1963 book has long been out of print. Thus, if you don't live near a major university library, you are probably out of luck if you want to read it— unless, of course, you visit Amazon, where, as of this writing, you can easily find five used copies priced as low as $25.15.

11. The original research paper is Alejandro Zentner, Michael D. Smith, and Cuneyd Kaya, "How Video Rental Patterns Change as Consumers Move Online," *Management Science* 59, no. 11 (2013): 2622–2634.

12. Erik Brynjolfsson, Yu (Jeffrey) Hu, and Duncan Simester, "Goodbye Pareto Principle, Hello Long Tail: The Effect of Search Costs on the Concentration of Product Sales," *Management Science* 57, no. 8 (2011): 1373–1386.

13. For more details, see Gal Oestreicher-Singer and Arun Sundararajan, "Recommendation Networks and the Long Tail of Electronic Commerce," *MIS Quarterly* 36, no. 1 (2012): 65–83.

14. As measured by the number of IMDb votes observed for the movie.

15. See Miguel Godinho de Matos, Pedro Ferreira, Michael D. Smith, and Rahul Telang, "Culling the Herd: Using Real World Randomized Experiments to Measure Social Bias with Known Costly Goods," *Management Science*, forthcoming.

16. For more details see Avi Goldfarb, Ryan C. McDevitt, Sampsa Samila, and Brian Silverman, "The Effect of Social Interaction on Economic Transactions: Evidence from Changes in Two Retail Formats," *Management Science*, forthcoming.

17. See https://hbr.org/2008/06/debating-the-long-tail and https://hbr.org/2008/07/the-long-tail-debate-a-response.

Chapter 6

1. Jeff Goodell, "Steve Jobs: The Rolling Stone Interview," *Rolling Stone*, December 3, 2003.

2. http://www.indiewire.com/article/guest-post-heres-how-piracy-hurts-indie-film-20140711

3. Music revenue in the United States fell from $14.6 billion in 1999 to $6.3 million in 2009. Source: http://money.cnn.com/2010/02/02/news/companies/napster_music _industry/

4. Source: Stan Liebowitz, "The Impacts of Internet Piracy," in *Handbook on the Economics of Copyright: A Guide for Students and Teachers*, ed. R. Watt (Edward Elgar, 2014).

5. For example, *MGM Studios v. Grokster*, a 2005 US Supreme Court decision that established "that one who distributes a device with the object of promoting its use to infringe copyright ... is liable for the resulting acts of infringement by third parties."

6. http://en.wikipedia.org/wiki/Stop_Online_Piracy_Act#cite_note-HousePress-28

7. https:/www.riaa.com/physicalpiracy.php?content_selector=piracy-online-scope -of-the-problem

8. http://ftp.jrc.es/EURdoc/JRC79605.pdf

9. http://www.cbc.ca/news/business/digital-piracy-not-harming-entertainment -industries-study-1.1894729

10. For a brief review of this literature with citations, see Michael Smith and Rahul Telang, "Competing with Free: The Impact of Movie Broadcasts on DVD Sales and Internet Piracy," *Management Information Systems Quarterly* 33, no. 2 (2009): 312–338.

11. Felix Oberholzer-Gee and Koleman Strumpf, "The Effect of File Sharing on Record Sales: An Empirical Analysis," *Journal of Political Economy* 115, no. 1 2007): 1–42.

12. Brett Danaher, Michael D. Smith, and Rahul Telang, "Piracy and Copyright Enforcement Mechanisms," in *Innovation Policy and the Economy*, volume 14, ed. J. Lerner and S. Stern (National Bureau of Economic Research, 2014).

13. Brett Danaher, Michael D. Smith, and Rahul Telang, "Copyright Enforcement in the Digital Age: Empirical Economic Evidence and Conclusions," prepared for tenth session of World Intellectual Property Organization Advisory Committee on Enforcement, Geneva.

14. The 2014 chapter included nineteen papers. The 2015 paper included two additional publications that appeared after the 2014 chapter went to press. The full list includes four additional papers that were brought to our attention after the 2015 paper became public.

15. Given this consensus, what should we make of the remaining three papers that found no evidence of harm? The most natural interpretation is that there are some settings in which piracy doesn't significantly harm sales. For example, table 6.1 in

the appendix to this chapter includes a paper in which we find no statistical harm from piracy that occurs at the time when a movie is shown on broadcast television networks (typically several years after a movie leaves the theaters), but we also note in the paper that these results "do not speak to the impact of piracy in the earlier part of a movie's life-cycle, where the availability of pirated content may have a negative impact on sales" (Michael Smith and Rahul Telang, "Competing with Free: The Impact of Movie Broadcasts on DVD Sales and Internet Piracy," *Management Information Systems Quarterly* 33, no. 2, 2009: 312–338, p. 336). It is also possible that the reported results are correct under the specific assumptions or empirical approach the authors used, but the results might change under a different set of assumptions or with a different empirical approach (see, for example, Rafael Rob and Joel Waldfogel, "Piracy on the High C's: Music Downloading, Sales Displacement, and Social Welfare in a Sample of College Students," *Journal of Law and Economics* 49, no. 1 (2006): 29–62; Stan Liebowitz, "How Reliable is the Oberholzer-Gee and Strumpf Paper on File-Sharing?" (http://ssrn.com/abstract=1014399); Stan Liebowitz, "The Oberholzer-Gee/Strumpf File-Sharing Instrument Fails the Laugh Test" (http://ssrn.com/abstract=1598037); George R. Barker and Tim J. Maloney, "The Impact of Free Music Downloads on the Purchase of Music CDs in Canada" (http://ssrn.com/abstract= 2128054)). In either case, the broader point is clear: In the vast majority of cases, piracy harms sales.

16. See, for example, Rob and Waldfogel, "Piracy on the High C's."

17. See http://www.ifpi.org/content/section_news/investing_in_music.html

18. See Joel Waldfogel, "Copyright Protection, Technological Change, and the Quality of New Products: Evidence from Recorded Music since Napster," *Journal of Law and Economics* 55 (2012), no. 4: 715–740.

19. Joel Waldfogel, "Copyright Protection, Technological Change, and the Quality of New Products: Evidence from Recorded Music since Napster," *Journal of Law and Economic* 55, no. 4 (2012): 715–740.

20. We observe similar results using other measures such as output measured by the Indian Censor Board. For a more complete discussion of our results, see Rahul Telang and Joel Waldfogel, "Piracy and New Product Creation: A Bollywood Story," 2014 (http://ssrn.com/abstract=2478755).

21. http://www.nytimes.com/2012/08/05/sunday-review/internet-pirates-will -always-win.html

22. http://www.bloomberg.com/bw/stories/1998-05-10/the-net-a-market-too -perfect-for-profits

23. See Michael Smith and Erik Brynjolfsson, "Customer Decision Making at an Internet Shopbot: Brand Still Matters," *Journal of Industrial Economics* 49, no. 4 (2001): 541–558.

24. Specifically, the control-group titles came from 53 shows on the CBS, CW, Fox, and NBC networks. Eighteen of these shows were available on Hulu before July 6 and experienced no change in availability in the four weeks after July 6, and 44 shows were not available on Hulu before July 6 and experienced no changes in availability in the four weeks after July 6. For a more detailed discussion, see Brett Danaher, Samita Dhanasobhon, Michael D. Smith, and Rahul Telang, "Economics of Digitization: An Agenda," in *Understanding Media Markets in the Digital Age: Economics and Methodology*, ed. A. Goldfarb, S. Greenstein, and C. Tucker (University of Chicago Press, 2015).

25. See Brett Danaher, Michael D. Smith, Rahul Telang, and Siwen Chen, "The Effect of Graduated Response Anti-Piracy Laws on Music Sales: Evidence from an Event Study in France," *Journal of Industrial Economics* 62, no. 3 (2014): 541–553.

26. Roger Parloff, "Megaupload and the Twilight of Copyright," *Fortune*, July 23, 2012: 21–24.

27. See Brett Danaher and Michael D. Smith, "Gone in 60 Seconds: The Impact of the Megaupload Shutdown on Movie Sales," *International Journal of Industrial Organization* 33 (2014), March: 1–8.

28. https://www.fbi.gov/news/pressrel/press-releases/justice-department-charges -leaders-of-megaupload-with-widespread-online-copyright-infringement

29. For more details of our approach, see Brett Danaher, Michael D. Smith, and Rahul Telang, The Effect of Piracy Website Blocking on Consumer Behavior, working paper, Carnegie Mellon University (available from http://ssrn.com/abstract =2612063).

Chapter 7

1. https://shotonwhat.com/cameras/canon-eos-5d-mark-iii-camera

2. The Academy Award for Best Editing went to *The Social Network* in 2010 and *The Girl with the Dragon Tattoo* in 2011. Previous nominees for this award that were edited with Final Cut Pro include *Cold Mountain* (2003), *No Country for Old Men* (2007), and *The Curious Case of Benjamin Button* (2008).

3. https://gigaom.com/2012/03/22/419-the-next-self-publishing-frontier-foreign -language-editions/

4. https://www.youtube.com/channel/UCy5mW8fB24ITiiC0etjLI6w

5. http://www.newyorker.com/magazine/2014/02/17/cheap-words

6. https://gigaom.com/2012/06/18/seth-godins-kickstarter-campaign-for-new-book -beats-40k-goal-in-3-5-hours/

7. https://www.kickstarter.com/projects/297519465/the-icarus-deception-why -make-new-from-seth-go

8. http://www.ew.com/article/2013/03/13/veronica-mars-movie-is-a-go-kickstarter

9. http://www.wsj.com/news/articles/SB1000142405270230363640457939732224 0026950

10. http://www.ew.com/article/2013/03/13/veronica-mars-movie-is-a-go-kickstarter

11. https://www.youtube.com/watch?v=CjW9I6jo7bQ

12. http://blogs.ocweekly.com/heardmentality/2014/05/nice_peter_epic_rap _battles_in_history.php

13. http://www.nytimes.com/2013/10/30/arts/television/epic-rap-battles-seeks -staying-power-on-youtube.html

14. http://www.statsheep.com/ERB

15. http://www.riaa.com/goldandplatinumdata.php?artist=%22Epic+Rap+Battles +of+History%22

16. ERB isn't the only YouTube success story. The most popular channel on You-Tube doesn't feature Katy Perry, Eminem, or Taylor Swift. It features Felix Kjellberg (a.k.a. PewDiePie), a 25-year-old man from Sweden who makes humorous videos of himself playing video games—and has 38 million subscribers and more than 9 billion video views worldwide. He also made an estimated $7 million in 2014 from his YouTube channel. (See http://www.bbc.com/news/technology-33425411.)

17. http://www.theguardian.com/books/2012/jan/12/amanda-hocking-self -publishing

18. http://www.deseretnews.com/article/865578461/Hip-hop-violinist-Lindsey -Stirling-overcomes-anorexia-critics-to-find-happiness-success.html

19. https://www.washingtonpost.com/blogs/the-switch/wp/2014/05/29/youtube -sensation-lindsey-stirling-on-how-the-internet-can-shape-the-music-industry/

20. https://www.youtube.com/user/lindseystomp/about

21. http://www.forbes.com/sites/michaelhumphrey/2011/10/26/epic-rap-battles -of-history-talking-brash-wit-with-a-youtube-hit/3/

22. http://www.billboard.com/articles/news/1559095/dubstep-violinist-lindsey -stirling-inks-deal-with-lady-gagas-manager

23. http://mediadecoder.blogs.nytimes.com/2011/03/24/self-publisher-signs-four -book-deal-with-macmillan/

24. http://content.time.com/time/arts/article/0,8599,1666973,00.html

25. http://www.wired.com/2007/12/ff-yorke/

26. https://louisck.net/news/a-statement-from-louis-c-k

27. https://louisck.net/news/another-statement-from-louis-c-k

28. http://recode.net/2015/01/31/louis-c-k-s-new-straight-to-fan-special-has-no -buzz-and-its-doing-better-than-his-first-one/

29. http://www.wired.com/2011/06/pottermore-details/

30. http://www.theguardian.com/books/booksblog/2012/mar/28/pottermore -ebook-amazon-harry-potter

31. http://nypost.com/2014/01/02/indie-artists-are-new-no-1-in-music-industry/

32. Joel Waldfogel and Imke Reimers, Storming the Gatekeepers: Digital Disinter- mediation in the Market for Books. working paper, University of Minnesota, Minneapolis, 2012.

33. http://www.washingtonpost.com/news/business/wp/2014/09/05/tv-is -increasingly-for-old-people/

34. http://www.dailymail.co.uk/news/article-2178341/Hollywood-Cinema -attendance-plummets-25-year-low.html

35. http://www.businessinsider.com/brutal-50-decline-in-tv-viewership-shows -why-your-cable-bill-is-so-high-2013-1

36. http://www.techhive.com/article/2833829/nearly-1-in-4-millennials-have-cut -the-cord-or-never-had-cable.html, cited by http://www.washingtonpost.com/news/ morning-mix/wp/2015/01/06/the-espn-streaming-deal-and-how-tv-is-becoming -entertainment-for-old-people/

37. http://blogs.wsj.com/cmo/2015/07/24/this-chart-shows-why-comcast-would-be -interested-in-vice-media-and-buzzfeed/

38. http://www.hollywoodreporter.com/news/study-5-percent-millennials-plan -732337

39. http://www.usatoday.com/story/tech/2014/12/19/youtube-diversity -millennials/18961677/

40. http://www.hollywoodreporter.com/news/study-5-percent-millennials -plan-732337

41. http://www.prnewswire.com/news-releases/sprint-and-suave-partner-with-leah -remini-to-create-consumer-generated-webisodes-58432852.html

42. http://www.nytimes.com/2009/03/25/arts/television/25moth.html?_r=1

43. http://adage.com/article/madisonvine-case-study/sprint-suave-find-success -mindshare-s-online-series/125090/

44. http://www.mediapost.com/publications/article/76165/suave-sprint-back-for -in-the-motherhood-webisod.html

45. http://variety.com/2008/scene/markets-festivals/abc-orders-motherhood -episodes-1117991763/

46. https://ewinsidetv.wordpress.com/2009/03/11/in-the-motherho/

47. Bowker, cited in Statistical Abstract of the United States: 2004–2005 (Government Printing Office), table 1129.

48. http://www.bowkerinfo.com/pubtrack/AnnualBookProduction2010/ISBN _Output_2002-2010.pdf. According to one recent estimate, on average Amazon adds a new book to its library every five minutes (http://techcrunch.com/2014/08/21/ there-is-one-new-book-on-amazon-every-five-minutes/).

49. http://www.musicsupervisor.com/just-how-many-releases-these-numbers -may-scare-you/

50. https://www.youtube.com/yt/press/statistics.html

Chapter 8

1. http://www.nytimes.com/2007/08/31/technology/31NBC.html

2. http://www.cnet.com/news/nbc-to-apple-build-antipiracy-into-itunes/

3. Philip Elmer-DeWitt, "NBC's Zucker: Apple Turned Dollars into Pennies," Fortune, October 29, 2007 (http://fortune.com/2007/10/29/nbcs-zucker-apple-turned-dollars -into-pennies/).

4. Quoted in "NBC Chief Warns Over iTunes Pricing," Financial Times, October 29, 2007 (http://www.ft.com/intl/cms/s/0/8f799be2-865a-11dc-b00e-0000779fd2ac .html).

5. See Brooks Barnes, "NBC Will Not Renew iTunes Contract," New York Times, August 31, 2007 (http://www.nytimes.com/2007/08/31/technology/31NBC.html). Apple estimated NBC's share of iTunes television sales at 30 percent (http://www .apple.com/pr/library/2007/08/31iTunes-Store-To-Stop-Selling-NBC-Television -Shows.html).

6. https://www.apple.com/pr/library/2007/09/05Apple-Unveils-iPod-touch.html

7. http://www.cnet.com/news/apple-slaps-back-at-nbc-in-itunes-spat/

8. http://www.nytimes.com/2007/09/20/business/media/20nbc.html

9. Target reportedly represented 15 percent of the DVD market (http://www.wsj
.com/articles/SB116035902475586468).

10. https://www.apple.com/pr/library/2006/09/12Apple-Announces-iTunes-7-with
-Amazing-New-Features.html

11. http://www.hollywoodreporter.com/news/target-blinks-dispute-disney-143682

12. Note that the vertical axis in figure 8.1 is on a log scale.

13. Brett Danaher, Samita Dhanasobhon, Michael D. Smith, and Rahul Telang,
"Converting Pirates without Cannibalizing Purchasers: The Impact of Digital Distribution on Physical Sales and Internet Piracy," *Marketing Science* 29, no. 6 (2010):
1138–1151.

14. As is documented in the journal article, we observed no increase in the availability of non-NBC content during this period, which suggested that the increased
availability of pirated episodes was causally related to the removal of NBC content
from iTunes.

15. One concession offered by Apple was two additional pricing points for content:
$0.99 for catalog content and $2.99 for high-definition episodes (http://www
.businessinsider.com/2008/9/nbc-s-zucker-we-came-back-to-itunes-because-we-got
-variable-pricing). But NBC received no concessions on piracy and no opportunity
to share in iPod revenue.

16. Brad Stone, *The Everything Store: Jeff Bezos and the Age of Amazon* (Little, Brown,
2013).

17. http://www.publishersweekly.com/pw/print/20040531/23431-amazon-co-op
-riles-independent-houses.html

18. http://www.newyorker.com/magazine/2014/02/17/cheap-words

19. Ibid.

20. http://www.publishersweekly.com/pw/print/20040531/23431-amazon-co-op
-riles-independent-houses.html

21. http://www.newyorker.com/magazine/2014/02/17/cheap-words

22. Quoted in Joe Miller, "Amazon Accused of 'Bullying' Smaller UK Publishers,"
BBC News, June 26, 2014. (http://www.bbc.com/news/technology-27994314).

23. http://articles.latimes.com/2011/oct/06/entertainment/la-et-jobs-music
-20111007

24. *Social Problems: Selections from CQ Researcher* (Pine Forge Press, 2009), p. 222.
See also http://featuresblogs.chicagotribune.com/entertainment_tv/2006/02/office
_workers.html.

25. http://www.newyorker.com/magazine/2014/02/03/outside-the-box-2

26. http://variety.com/2009/digital/features/online-distribution-pulls-ahead-of-film -111799As9758/

27. http://www.digitalbookworld.com/2013/e-retailers-now-accounting-for-nearly -half-of-book-purchases-by-volume/

28. http://www.theverge.com/2015/4/15/8419567/digital-physical-music-sales -overtake-globally

29. http://partners.nytimes.com/library/tech/99/03/biztech/articles/14amazon .html

30. Michael Smith, Joseph Bailey, and Erik Brynjolfsson, "Understanding Digital Markets: Review and Assessment," in *Understanding the Digital Economy*, ed. E. Brynjolfsson and B. Kahin (MIT Press, 2000).

31. E. J. Johnson, S. Bellman, and G. L. Lohse, "Cognitive Lock-in and the Power Law Of Practice," *Journal of Marketing* 67, no. 2 (2002): 62–75.

32. For example, Erik Brynjolfsson, Astrid Dick, and Michael Smith analyzed data from an online price-comparison site and found that these price-conscious consumers almost never searched to the second page of prices, even though the lower pages included deals that, on average, provided $6 more value to the consumer than the best offer on the first page of prices ("A Nearly Perfect Market? Differentiation Versus Price in Consumer Choice," *Quantitative Marketing and Economics* 8, no.1 (2010): 1–3). In effect, consumers were willing to forgo $6 worth of value in exchange for not having to invest the time and cognitive effort necessary to process additional product choices. This result is consistent with related research showing that online consumers face similarly high costs from simple tasks such as entering eBay auctions (P. Bajari and A. Hortaçsu, "The Winner's Curse, Reserve Prices, and Endogenous Entry: Empirical Insights from eBay Auctions," *RAND Journal of Economics* 34 (2003): 329–355), bidding on online auctions (I. Hann and C. Terwiesch, "Measuring the Frictional Cost of Online Transactions: The Case of a Name-Your-Own-Price Channel," *Management Science* 49 (2003): 1563–1579), and searching for textbooks (H. Hong and M. Shum, "Using price distributions to estimate search costs," *RAND Journal of Economics* 37 (2006): 257–275).

33. This isn't an argument against the use of Digital Rights Management software per se. The data suggest that DRM can reduce harm from casual piracy in some settings. (See, for example, Imke Reimers' finding that the use of DRM protection increases e-book sales by 15.4 percent, at http://www.econ.umn.edu/~reime062/ research/piracy_paper.pdf.) Rather, it is an argument that these advantages should be weighed against the disadvantages associated with platform lock-in.

34. See, for example, Nicola F. Sharpe and Olufunmilayo B. Arewa, "Is Apple Playing Fair? Navigating the iPod FairPlay DRM Controversy," *Northwestern Journal of*

Technology and Intellectual Property 5, no. 2: 331–349; Herbert Hovenkamp, Mark D. Janis, Mark A Lemley, and Christopher R. Leslie, *IP and Antitrust: An Analysis of Antitrust Principles Applied to Intellectual Property Law*, second edition (Wolters Kluwer Law & Business, 2014; Thorsten Kaseberg, *Intellectual Property, Antitrust and Cumulative Innovation in the EU and the US* (Bloomsbury, 2012).

35. Yannis Bakos and Erik Brynjolfsson, "Bundling and Competition on the Internet," *Marketing Science* 19, no. 1 (2000): 63–82.

36. Ibid.

37. http://arstechnica.com/uncategorized/2007/11/hands-on-nbc-direct-beta-makes -hulu-seem-utopian-not-ready-for-beta-tag/

38. http://fortune.com/2014/12/09/hbo-streaming/

39. http://variety.com/2014/digital/news/hbo-cto-otto-berkes-resigns-as -network-enlists-mlb-to-build-ott-platform-1201375255/

Chapter 9

1. Michael Lewis, *Moneyball* (Norton, 2003), pp. 219–220.

2. Ibid., p. 233.

3. Ibid., p. 57.

4. Ibid.

5. http://www.newyorker.com/magazine/2014/02/03/outside-the-box-2

6. http://www.newyorker.com/magazine/2014/02/17/cheap-words

7. Ken Auletta, "Publish or Perish," *The New Yorker*, April 26, 2010.

8. http://www.newyorker.com/reporting/2014/02/17/140217fa_fact_packer

9. http://www.hollywoodreporter.com/news/sonys-michael-lynton-defends-studio -759494

10. http://www.nytimes.com/2013/02/25/business/media/for-house-of-cards-using -big-data-to-guarantee-its-popularity.html?_r=1

11. As Pittsburgh Pirates fans, we are bitter about this.

12. We have heard of some instances in which platform companies have been willing to share more detailed data on customers or to facilitate direct marketing programs, but only under special circumstances or for a fee. This reinforces two larger points we have been making: that customer data and customer access have become important strategic assets, and that the ability to control these assets gives large distributors significant leverage in negotiations with their partners.

13. http://www.hollywoodreporter.com/news/aftermath-hulu-ceos-bad-boy-101517

14. See, for example, http://variety.com/2014/digital/news/amazon-to-spend-more
-than-100-million-on-original-series-in-q3-1201268987/, http://variety.com/2015/
digital/news/amazon-studios-to-produce-movies-for-theatrical-digital-release-
in-2015-1201408688/, and http://www.wsj.com/articles/youtube-seeks-streaming
-right-to-tv-shows-movies-1449104356

15. http://youtube-global.blogspot.com/2015/10/red-originals.html

16. http://www.vulture.com/2015/07/netflix-original-programming-hbo-fx.html

17. Gina Keating, *Netflixed: The Epic Battle for America's Eyeballs* (Portfolio, 2013).

18. http://www.nytimes.com/2013/02/25/business/media/for-house-of-cards-using-
big-data-to-guarantee-its-popularity.html

19. http://www.hollywoodreporter.com/news/amazon-studios-head-roy-price
-721867

20. John Seabrook, "Revenue Streams," *The New Yorker*, November 24, 2014 (http://
www.newyorker.com/magazine/2014/11/24/revenue-streams)

21. http://www.theatlantic.com/magazine/archive/2014/12/the-shazam-effect/
382237/

22. http://www.newyorker.com/reporting/2014/02/17/140217fa_fact_packer

23. "A Chat with Ted Sarandos, Mitch Hurwitz, and Vince Gilligan," National Association of Television Program Executives, January 21, 2015 (https://www.youtube
.com/watch?v=Zdy8-FDV7c0).

24. Kevin Spacey, keynote address, Content Marketing World 2014, Cleveland,
September 11, 2014.

25. Source: http://variety.com/2015/tv/news/golden-globe-nominations-2016-hbo
-nbc-1201658385/

26. Source: http://deadline.com/2015/12/golden-globes-nominations-2016-tv-series
-networks-list-1201664377/

27. See, for example, http://www.hollywoodreporter.com/news/breaking-bad-how
-cable-netflix-619857.

28. David Bank of RBC Capital Markets predicted that in 2015 the networks and
studios would receive $6.8 billion from Netflix, Hulu, and Amazon for the rights
to stream television reruns (http://www.wsj.com/articles/netflix-viewership-finally
-gets-a-yardstick-1440630513).

Chapter 10

1. Quoted in "How to Survive in Vegas," *Business Week*, August 9, 2010 (http://www.bloomberg.com/bw/magazine/content/10_33/b4191070705858.htm).

2. Our telling of the Harrah's story relies heavily on the following three sources: Rajiv Lal, Harrah's Entertainment, case study, Harvard Business School, 2002; Victoria Chang and Jeffrey Pfeffer, Case OB-45, Gary Loveman and Harrah's Entertainment, Stanford Graduate School of Business, 2003; Gary Loveman, "Diamonds in the Data Mine," *Harvard Business Review*, May 2003.

3. Rajiv Lal and Patricia Carrolo, Harrah's Entertainment Inc., case 502-011, Harvard Business School, 2001, p. 3.

4. Ibid.

5. Ibid., p. 5.

6. Loveman, "Diamonds in the Data Mine," p. 4.

7. Ibid.

8. Lal and Carrolo, Harrah's Entertainment Inc., p. 6.

9. Chang and Pfeffer, Gary Loveman and Harrah's Entertainment.

10. Richard Metters, Carrie Queenan, Mark Ferguson, Laura Harrison, Jon Higbie, Stan Ward, Bruce Barfield, Tammy Farley, H. Ahmet Kuyumcu, and Amar Duggasani, "The 'Killer Application' of Revenue Management: Harrah's Cherokee Casino and Hotel," *Interfaces* 38, no. 3 (2008): 161–175.

11. Loveman, "Diamonds in the Data Mine," p. 4.

12. Chang and Pfeffer, Gary Loveman and Harrah's Entertainment.

13. Meridith Levinson, "Harrah's Knows What You Did Last Night," CIO Newsletter, June 6, 2001 (http://www.cio.com.au/article/44514/harrah_knows_what_did_last_night/).

14. Chang and Pfeffer, Gary Loveman and Harrah's Entertainment.

15. Richard H. Levey, "Destination Anywhere: Harrah's Entertainment Inc.'s Marketing Strategy," *Direct*, 1999, cited in Lal and Carrolo, Harrah's Entertainment Inc.

16. Loveman, "Diamonds in the Data Mine," p. 3.

17. Ibid., p. 4.

18. Gary Loveman in the Gaming Hall of Fame for 2013, Gambling USA, September 14, 2013 (http://www.gamblingusa.com/gary-loveman-gaming-hall-fame-2013/).

19. Kate O'Keeffe, "Real Prize in Caesars Fight: Data on Players," *Wall Street Journal*, March 19, 2015 (http://www.wsj.com/articles/in-caesars-fight-data-on-players-is-real -prize-1426800166).

20. Steve Knopper, *Appetite for Self-Destruction: The Spectacular Crash of the Record Industry in the Digital Age* (Simon and Schuster, 2009).

21. Source: IFPI, "Music industry revenue worldwide from 2002 to 2014, by sector (in billion U.S. dollars)" (http://www.statista.com/statistics/272306/worldwide -revenues-of-the-music-industry-by-category/).

22. We describe the high-level results of our study below. For further details, see Brett Danaher, Yan Huang, Michael D. Smith, and Rahul Telang, "An Empirical Analysis of Digital Music Bundling Strategies," *Management Science* 60, no. 9 (2015): 1413–1433.

23. http://www.nielsen.com/us/en/insights/reports/2015/the-total-audience -report-q1-2015.html

24. http://www.wsj.com/articles/viacom-beats-expectations-on-ninja-turtles -transformers-1415881443

25. http://blogs.wsj.com/cmo/2015/06/25/nielsen-mitch-barns-tv-networks-netflix/

26. See Filipa Reis, Miguel Godinho de Matos, and Pedro Ferreira, The Impact of Convergence Technologies on the Substitution Between TV and Internet: Evidence from a Randomized Field Experiment, working paper, Carnegie Mellon University, 2015. In a separate test, Reis et al. analyzed Internet use among consumers who were given access to premium television channels but not the DVR capabilities; they found no change in Internet consumption. That finding, they argue, suggests that "when the TV provides an experience similar to that of video streaming on the Internet, users watch more of it."

27. For a more detailed discussion of our experimental approach, see Jing Gong, Michael D. Smith, and Rahul Telang, "Substitution or Promotion? The Impact of Price Discounts on Cross-Channel Sales of Digital Movies," *Journal of Retailing* 91, no. 2 (2015): 343–357.

28. These mass-market strategies can be effective, of course. For example one of the most important mass-market advertising events is the Super Bowl. In a recent work- ing paper titled "Super Returns to Super Bowl Ads?" (http://people.ischool.berkeley .edu/~hal/Papers/2015/super.pdf), Seth Stephens-Davidowitz, Hal Varian, and Michael D. Smith analyzed the effectiveness of Super Bowl advertising on movie revenue. Not surprisingly, they weren't able to run an experiment per se—but they were able to analyze the results of a series of natural experiments. Their analysis relied on the fact that cities whose home teams are in the Super Bowl have much larger audiences for the game than other cities and on the fact that Super Bowl ads

are purchased well before anyone knows which teams will play in the game. The increased viewership in cities whose teams are playing in the game acts as an exogenous shock to the number of people who see the advertisements. That allowed Stephens-Davidowitz et al. to analyze theatrical attendance for all 54 movies that were advertised during Super Bowl broadcasts from 2004 to 2012. They found that many more people watched these movies in the cities whose teams played in the game and that, in terms of return on investment, a $3 million Super Bowl advertisement for a movie generated, on average, $8.4 million in increased profits to the studio.

Chapter 11

1. John Markoff, "Michael Dell Should Eat His Words, Apple Chief Suggests," *New York Times*, January 16, 2006 (http://www.nytimes.com/2006/01/16/technology/16apple.html).

2. http://www.cnet.com/news/gateway-shuts-10-percent-of-u-s-stores/.   Gateway would close its remaining 188 stores in April of 2004 (http://www.pcworld.com/article/115507/article.html).

3. http://www.bloomberg.com/bw/stories/2001-05-20/commentary-sorry-steve -heres-why-apple-stores-wont-work

4. http://www.forbes.com/sites/carminegallo/2015/04/08/why-the-experts-failed-to -predict-the-apple-stores-success/

5. http://fortune.com/2015/03/13/apples-holiday-top-10-retailers-iphone/

6. http://www.forbes.com/sites/carminegallo/2015/04/08/why-the-experts-failed-to -predict-the-apple-stores-success/

7. http://fortune.com/2011/08/26/how-apple-became-the-best-retailer-in-america/

8. http://bits.blogs.nytimes.com/2011/11/25/a-look-at-apples-spot-the-shopper -technology/

9. Source: http://variety.com/2015/film/news/godzilla-vs-king-kong-legendary-ceo -1201656742/

10. http://ir.aol.com/phoenix.zhtml?c=147895&p=irol-newsArticle_print&ID =1354531

11. http://variety.com/2015/digital/news/netflix-bandwidth-usage-internet-traffic -1201507187/

12. http://fortune.com/2012/08/20/hulus-network-drama/

13. Jason Kilar, "Stewart, Colbert, and Hulu's Thoughts about the Future of TV," http://blog.hulu.com/2011/02/02/stewart-colbert-and-hulus-thoughts-about -the-future-of-tv/

14. http://allthingsd.com/20110203/is-jason-kilar-trying-to-get-fired/

15. http://www.ft.com/intl/cms/s/0/2503f886-2f60-11e0-834f-00144feabdc0.html

16. Ibid.

17. Janet Morrissey, "The Beginning of the End for Hulu?" *Fortune*, January 8, 2013.

18. For a more details on the methods and results, see Brett Danaher, Michael D. Smith, and Rahul Telang, Windows of Opportunity: The Impact of Early Digital Movie Releases in the Home Entertainment Window, working paper, Carnegie Mellon University, 2015.

19. One of the earliest academic studies of direct consumer observation was William D. Wells and Leonard A. Lo Sciuto, "Direct Observation of Purchasing Behavior," *Journal of Marketing Research* 3, no. 3 (1966): 227–233. Wells and Lo Sciuto reported having data collectors follow consumers in grocery stores and record their behavior in detail. Their method required 600 hours of labor on the part of the collectors to obtain data from 1,500 shopping episodes.

20. Peter E. Rossi, Robert E. McCulloch, and Greg M. Allenby, "The Value of Purchase History Data in Target Marketing," *Marketing Science* 15, no. 4 (1996): 321–340.

21. Food Marketing Institute, Variety of Duplication: A Process to Know Where You Stand. Prepared by Willard Bishop Consulting and Information Resources, Inc. in cooperation with Frito-Lay, 1993.

22. Robert D. Austin and Warren McFarlan, H. E. Butt Grocery Company: A Leader in ECR Implementation (B) (Abridged), case 9-198-016, Harvard Business School, 1997, p. 2.

Index